Lecture Notes
in Control and Information Sciences

216

Editor: M. Thoma

W0042600

Springer-Verlag London Ltd.

Rudolf Kulhavý

Recursive Nonlinear Estimation

A Geometric Approach

Springer

Author

Dr Rudolf Kulhavý
Institute of Information Theory and Automation
Academy of Sciences of the Czech Republic
P.O. Box 18, 182 08 Prague, Czech Republic

ISBN 978-3-540-76063-4

British Library Cataloguing in Publication Data
Kulhavý, Rudolf
  Recursive nonlinear estimation : a geometric approach. -
  (Lecture notes in control and information sciences ; 216)
  1.Parameter estimation 2.Estimation theory
  I.Title
  519.5'4
ISBN 978-3-540-76063-4   ISBN 978-3-540-40947-2 (eBook)
DOI 10.1007/978-3-540-40947-2

Library of Congress Cataloging-in-Publication Data
A catalog record for this book is available from the Library of Congress

Typesetting: Camera ready by author

69/3830-543210 Printed on acid-free paper

To Lenka, Lukáš and Martin

Preface

Motivation

The theory of parameter estimation is known to be a well-developed, mature field of applied mathematics. Some of the probabilistic tools commonly utilized now in statistics originated in the late 17th and early 18th century. Yet, solutions optimal from the theoretical point of view may be of little practical value if they involve computations that are too difficult to be implemented on any real computer. This is typically the case when the models considered are non-linear, non-normal, or high-dimensional. There are essentially two ways of arriving at a feasible algorithm—either to look for a simpler model, or to approximate the theoretical solution. In both cases, the ideal solution is abandoned.

We clearly face a dilemma. Should we stick to theory—and compute little? Or should we rather insist on feasibility—and sacrifice theory? This book suggests a compromising solution that views approximation as an inherent part of inference. The bottom line is that no substantial difference is made between the *statistical* and *computational* uncertainty.

Level

The book is intended for scientists and graduate students interested in parameter estimation and system identification for complex, 'nonstandard' models. The mathematical prerequisites are rather low although some preliminary knowledge of statistical methods of parameter estimation will be helpful.

Every attempt was made to avoid unnecessary technicalities. This is consistent with the author's firm belief that an escape from the dilemma 'statistical theory versus feasible algorithms' will eventually be found in a thoughtful modification of the current paradigms rather than in higher mathematics itself.

Throughout the book it is assumed that data can be modelled as either continuous or discrete random variables. The density functions are considered strictly positive. The unknown parameters are supposed to come from a subset of the real space. These restrictions allow us to avoid the systematic use of measure theory and introduction of differential-geometric notions. The loss of generality is, hopefully, outweighed by a more lucid and widely acceptable presentation, with focus on ideas rather than technical points.

Organization

The book is organized as a step-by-step introduction into a different way of thinking about statistical methods of parameter estimation, characterized by the systematic use of information "distances" rather than the likelihood function or posterior density of the unknown parameter.

Chapter 1 introduces the reader into the problem, explains why the theoretical paradigms need to be approximated and gives a brief survey of typical approaches to design of approximation.

In Chap. 2, the problem of probability-based parameter estimation is transposed into the language of information measures. The likelihood function and the posterior density are shown to be directly related to the classical information measures—inaccuracy, Kullback-Leibler distance and Shannon's entropy. A number of examples illustrate the view of parameter estimation as 'probability matching'.

Chapter 3 introduces the key tool which is decomposition of the information "distance" between the empirical and theoretical densities into two components—one independent of data given the value of a properly chosen statistic, the other independent of the unknown parameter. It gives rise to a specific Pythagorean-like geometry of probability density functions. The connection between the Pythagorean-like decomposition and the statistical notion of a sufficient statistic is elucidated.

Chapter 4 applies the Pythagorean geometry for approximation of the likelihood function and posterior density in the case that the statistic used is not sufficient for the model considered. A proper choice of the statistic is discussed in detail. The construction proposed is shown to be equivalent to the choice of a necessary statistic.

Chapter 5 discusses some of possible ways of numerical implementation of the approximate estimation. A simple modification of the basic scheme is shown that makes the basic algorithm capable of tracking slowly varying parameters. The effect of the modification on the posterior density is essentially equivalent with the effect of exponential forgetting.

Chapter 6 gives a summary of key points of the proposed solution, with emphasis on the features that distinguish it from other approaches. Open problems and challenges are indicated.

Throughout the book the cases of independent observations and controlled dynamic systems are considered in parallel. The former, simpler and more transparent case serves as introduction and motivation to the latter which is the real challenge.

Geometric Insight

I build as much as possible on geometric intuition, making maximum use of orthogonal and minimum distance projections. I like to envisage key ideas through pictures and geometric concepts and I do hope that most readers will enjoy it too. The Euclidean case, simple and transparent, serves repetitively as motivation for analogous solutions in the spaces of probability density functions. A number of examples and solved problems is used to illustrate and enlighten the major steps of exposition.

Acknowledgments

The book has been greatly influenced, both in contents and form, by numerous discussions with colleagues and friends working in fields as different as system identification, statistical parameter estimation, differential geometry, probability theory, information theory, mathematical system theory. The list of their names would be too long to put here, but my thanks go to all of them for making me see the problem in a broader perspective.

My personal enthusiasm for the geometric way of thinking about statistical problems started in 1985 when I discovered in one of Moscow's second-hand bookshops the monograph *Statistical Decision Rules and Optimal Inference* by Nikolaĭ N. Čencov. I take this opportunity to pay tribute to the book that has become a source of unceasing inspiration for me.

Results achieved since 1985 have been to a large extent the product of a stimulating environment of the Institute of Information Theory and Automation in Prague. My colleagues and students contributed a good deal to the production of this book, both by raising tough questions and by helping me to see the answers. I am particularly indebted to Václav Peterka who introduced me into the vast gardens of Bayesian identification and to Miroslav Kárný who roused my interest in recursive nonlinear estimation and who constantly encouraged and supported me on my way through.

The biggest debt I owe is to my family—for all those hours that ought to have been spent with them.

Prague, March 1996   Rudolf Kulhavý

Table of Contents

Basic Notations and Conventions

Special Symbols

$\triangleq$	defined as
$\propto$	proportionality, i.e., equality up to a normalizing factor
$\square$	end of example

Sets

$\mathcal{X} = \{x_1, \ldots, x_n\}$	$\mathcal{X}$ is a set having elements $x_1, \ldots, x_n$
$a \in \mathcal{X}$	a is an element of $\mathcal{X}$
$A \subset \mathcal{X}$	A is a subset of $\mathcal{X}$
$|\mathcal{X}|$	number of elements of the set $\mathcal{X}$
$\emptyset$	empty set
$\mathbb{R}$	set of real numbers

Cartesian Products

$\mathcal{Y} \times \mathcal{Z}$	Cartesian product of the sets $\mathcal{Y}$ and $\mathcal{Z}$
$(y, z) \in \mathcal{Y} \times \mathcal{Z}$	(y, z) is an element of $\mathcal{Y} \times \mathcal{Z}$

Vectors and Sequences

v^T	vector transposed to the vector v
$\|v\|$	Euclidean norm of the vector v, $\|v\|^2 = \sum_{i=1}^{n} v_i^2$
$y^N = (y_1, \ldots, y_N)$	a finite sequence (row vector) of length N starting at 1

$y_{m+1}^{N+m} = (y_{m+1}, \ldots, y_{N+m})$ a finite sequence (row vector) of length N starting at $m+1$

$\mathcal{Y}^N$ N-th Cartesian power of $\mathcal{Y}$, i.e., the set of sequences of length N of elements of $\mathcal{Y}$

$y^N \in \mathcal{Y}^N$ y^N is an element of $\mathcal{Y}^N$

Functions

$f: \mathcal{Y} \to \mathbb{R}$ mapping of $\mathcal{Y}$ into $\mathbb{R}$

$f^{-1}(y) = \{x : f(x) = y\}$ the inverse image of y

$\exp(x)$ exponential of x

$\log(x)$ natural logarithm of x

Gradient and Hessian

$\nabla_\lambda f(\lambda)$ gradient of the function $f: \mathbb{R}^n \to \mathbb{R}$, i.e., a column vector of first-order derivatives $\frac{\partial}{\partial \lambda_i} f(\lambda)$

$\nabla_\lambda^2 f(\lambda)$ Hessian of the function $f: \mathbb{R}^n \to \mathbb{R}$, i.e., a square symmetric matrix of second-order derivatives $\frac{\partial^2}{\partial \lambda_i \partial \lambda_j} f(\lambda)$

Random Variables

$(\Omega, \mathscr{F}, \mu)$ probability space with Ω being a set, $\mathscr{F}$ a σ-algebra of its subsets and μ a probability measure on $\mathscr{F}$

$Y: \Omega \to \mathcal{Y}$ random variable; if $\mathcal{Y}$ is finite, $Y^{-1}(y) \in \mathscr{F}$ for every $y \in \mathcal{Y}$

$\Pr\{Y \in A\}$ probability that the random variable Y takes value in the set $A \subset \Omega$, being the μ-measure $\mu(\{\omega : Y(\omega) \in A\})$

$Y^N = (Y_1, \ldots, Y_N)$ finite sequence (row vector) of random variables

Random and Observed Samples

Y_k, y_k autonomous signal or output of system

U_k, u_k directly manipulated input to system

Z_k, z_k known vector function of past outputs Y^{k-1} and past inputs U^k

Probability Density Functions

$r_N(y)$ empirical density of the random variable Y

$s_\theta(y)$ theoretical density of the random variable Y parametrized by the parameter $\theta \in \mathcal{T}$

$q_\theta^N(y^N)$ joint density of the sample Y^N parametrized by $\theta \in \mathcal{T}$

$r_N(y,z)$ joint empirical density of the random variables (Y,Z)

$\bar{r}_N(z)$ marginal density of the random variable Z

$s_\theta(y|z)$ conditional theoretical density of the random variable Y given another variable Z, parametrized by $\theta \in \mathcal{T}$

$q_\theta^N(y^{N+m}, u^{N+m}|y^m, u^m)$ joint density of the sample $Y_{m+1}^{N+m}, U_{m+1}^{N+m}$ conditional on initial values, parametrized by $\theta \in \mathcal{T}$

$l_N(\theta)$ likelihood function, i.e., joint density of N observed data taken as a function of the parameter θ

$p_N(\theta)$ posterior density of the parameter Θ conditional on N observed data

Expectations

$E_N(h(Y))$ empirical expectation of the random variable $h(Y)$ with respect to the empirical density $r_N(y)$

$\mathrm{Var}_N(h(Y))$ empirical variance of the random variable $h(Y)$ with respect to the empirical density $r_N(y)$

$E_\theta(h(Y))$ theoretical expectation of the random variable $h(Y)$ with respect to the density $s_\theta(y)$

Independence

$X \perp Y$ X and Y are independent random variables

$X \perp Y | Z$ X and Y are conditionally independent given $Z = z$

Information Measures

$K(r{:}s)$	inaccuracy of the density $r(y)$ relative to the density $s(y)$
$D(r\|s)$	Kullback-Leibler distance of the densities $r(y)$ and $s(y)$
$H(r)$	Shannon's entropy of the density $r(y)$
$\bar{K}(r{:}s)$	conditional inaccuracy of the joint density $r(y,z)$ relative to the conditional density $s(y\|z)$
$\bar{D}(r{:}s)$	conditional Kullback-Leibler distance between the joint density $r(y,z)$ and the conditional density $s(y\|z)$
$\bar{D}(s{:}s'\|\bar{r})$	conditional Kullback-Leibler distance between the conditional densities $s(y\|z)$ and $s'(y\|z)$ given the marginal density $\bar{r}(z)$

Data Statistic

$h{:}\mathcal{Y} \to \mathbb{R}^n$	single-data statistic, i.e., a finite-dimensional mapping of single observation
$h{:}\mathcal{Y} \times \mathcal{Z} \to \mathbb{R}^n$	single-data statistic for dependent data
$\bar{h}_N$	empirical expectation of $h(Y)$ or $h(Y,Z)$

1. Inference Under Constraints

As the engineering science is confronted with increasingly difficult problems, the limits of current paradigms become more tangible. The use of probability-based methods of parameter estimation for complex models is no exception in this respect. The conflict between the abstract mathematical theory and finite computer resources necessarily calls for approximation of the theoretically optimal solutions. The problem the user usually faces is the abundance of possible approximations, shortage of guidelines as to when to use a particular one and little information on how far the approximation is from the optimum.

The area of approximate estimation is vast and fuzzy; this chapter gives only a brief survey of common approaches to approximation of probability-based estimation. The stress is put on Bayesian methods and the problem of approximate computation of the likelihood function and posterior density.

1.1 The Bayesian Paradigm

It sounds quite natural to describe the random behaviour of data through their probability distribution. It is possible though not so natural to describe even the uncertainty of assigning such a distribution in terms of probability. Probability can thus appear in a double role. On the one hand, as a relative frequency of possible outcomes of a certain experiment in a long run. On the other hand, as a measure of uncertainty of unknown quantities.

Jacob Bernoulli, Thomas Bayes and Pierre Simon Laplace came with the idea of 'inverse probability' as early as in the 18th century. Much later, in this century, Keynes (1921), de Finetti (1937), Jeffreys (1939), Cox (1946), Savage (1954), de Finetti (1974) showed that the rules of consistent reasoning are essentially the laws of probability theory.

The symmetric view of probability, usually ascribed to T. Bayes, has turned out useful in many areas. It has brought some unique features into *statistics* forming thus its self-contained branch. It has put *artificial intelligence* on a firm basis and set up a reference solution for design of *expert systems*. It has been applied with success in *physics*, *engineering* and *statistics* to cope with ill-posed (underdetermined) problems that commonly appear in problems like identification, fault detection, econometrics, control, medical diagnosis, geophysical exploration, image processing, synthesis of electrical filters or optical systems.

The pros and cons of the Bayesian paradigm can be well illustrated on the case of stochastic adaptive control. Suppose we are to control a dynamic system that depends on some unknown parameter θ. We have two options then. Either we satisfy with a point estimate of θ, or we take the uncertainty of θ into account. In the latter case, the expectation in a cost function applies to both the stochastic behaviour of the system and the uncertainty of the parameter θ. This converts the original problem into a hyperproblem that has no more unknowns. Its *information state* is formed by the probability density function of the original state and the parameter θ conditional on the observed data.

The beauty of the result is that it looks as if all uncertainty vanished. As soon as the prior density is set, the state evolves in a definite way, governed by the laws of probability theory. The appeal of the solution is paid, however, by the immense dimension of the information state. Unless the problem has a finite-dimensional statistic, there is no feasible way of updating the full information state. The reader interested in detail can consult, e.g., Bertsekas and Shreve (1978), Kumar (1985), Kárný et al. (1985) or Kumar and Varaiya (1986).

The above limitation is an inherent difficulty of the Bayesian inference that is not bound to just the control problem. In fact, it is felt urgently in any of the above mentioned application areas. But, when the ideal paradigm cannot be followed, what is one to substitute for it? Can we cut back the extreme dimension of the theoretical optimum and yet stay close to it? Can approximate estimation keep something from the logical soundness of the original scheme? These challenging questions form the main theme of the book.

1.2 Recursive Estimation

The way the Bayesian estimation is implemented is largely affected by the amount of data processed.

One extreme arises when we are given a complete sample that is composed of relatively *few data*. Obviously, a limited amount of data needs only a limited computer memory. Moreover, data can be processed off-line so that the total time of computation is only weakly limited. Thus, in this case we can stay very close to the ideal solution. The reason why we cannot perhaps reach the optimum is in errors we make in numerical integration and optimization. Both operations are often called for—to compute mean values, to calculate marginal densities, to maximize probability. The complexity of integration and optimization grows quickly with the dimension of the underlying spaces. In high dimension, special methods are necessary for their effective solutions (see Sect. 1.4).

The opposite extreme occurs when the sample is composed of *many data*. It is not difficult to see that the large amount of data makes estimation easier—stressing the regular pattern in the stochastic behaviour of data and reducing the uncertainty of the unknown parameters. As a result, we usually need only the resulting estimates and need not bother about the transient modes. The theory of point estimation focuses on analysing the asymptotic properties of such estimators—their consistency, efficiency, approximate normality, rate of convergence etc.

Somewhere between the two extremes lies the case when there are too many data to carry out full analysis but too few to rely on asymptotics. This happens namely when the following two points apply together:

(a) Parameter estimation is required to supply, at regular time instants, input information for subsequent decision-making such as optimal control or signal processing. The estimation algorithm must therefore be able to process data sequentially as they are observed.

(b) The model describes the actual process behaviour only locally. The model parameters thus depend on the current state of the process and vary in time. The amount of data relevant for the current working point is typically limited. To prevent mixing up relevant and obsolete data in parameter estimation, we must introduce an explicit model of parameter variations or apply some heuristics how to "forget" obsolete information.

The recursive character of computations calls for massive compression of data. This together with the moderate amount of data available for estimation produces significant posterior uncertainty of the unknown parameters which cannot be neglected or easily approximated. Both the features make approximation of the Bayesian estimation a delicate matter where one must really care how far he is from the theoretical paradigm.

The last case is of primary interest for us in the book. Note that estimation with the adjectives *adaptive* and *recursive* is prerequisite for solution of many problems that engineering science deals with, typically coming under the headings of adaptive control (cf. Fig. 1.1) or adaptive signal processing. It has become a challenging topic in statistics and econometrics as well.

1.3 Admissible Complexity

Computer resources are always limited and any numerical computation must be accomplished in a given space, time and precision. The limited resources often make the ideal inference impracticable. Then either the model, or the estimation algorithm must be modified so to meet given constraints.

In the former case, we hope to find another model that is feasibly identifiable and still gives a sufficiently accurate description of the actual behaviour. Reducing, transforming, decomposing the original model may suggest a suitable approximation. This approach is used very often in practice and it may be quite successful.

When the cost for simplification is too high, we have to undertake the latter, more risky way which is approximation of the optimal estimation algorithm. The danger is obvious: any approximation sets up essentially a new paradigm of inference different from the theoretically optimal one. The bad thing is that the new "paradigm" arises often *ad hoc*. We gain feasibility, but lose a sound theoretical background.

A purely numerical approach to approximation neglects that the uncertainty of estimation is affected by the imprecision of its computational implementation. Clearly,

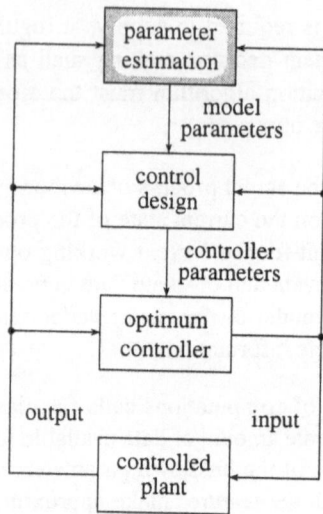

Fig. 1.1. A common structure of the self-tuning controller.

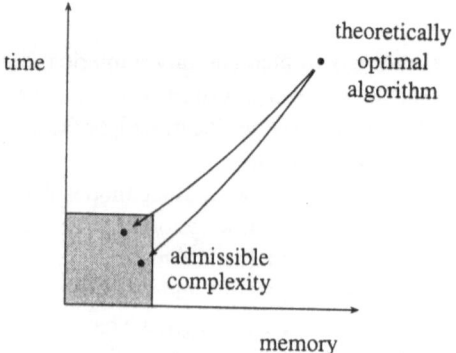

Fig. 1.2. Ambiguity of feasible approximation.

the farther the actual estimator is from the optimal one, the more uncertain the inference is likely to be (for finite samples, at least). An adequate expression of this seems to be missing in theory.

Conceptually at least, a simple solution exists. Its bottom line is to admit that there is no substantial difference between the pieces of uncertainty born in estimation and computation. Both kinds of uncertainty deserve to be treated equally and put together so to comply with common inference rules. Then the decrease of complexity is directly reflected by the increase of uncertainty of estimates. However natural the idea sounds, it is too vague in this raw form to be applied right away. We need to translate it first into terms that are compatible with the statistical framework used.

Strictly speaking, the feasibility problem we speak about is rather a collection of problems related to both data compression and computational complexity. It comes as no surprise that particular facets of the feasibility problem can be translated into the statistical language with variable success. In the following chapters we focus mainly on problems related to the loss of information due to compression of data.

Let us add for the sake of completeness that the concept of complexity is well established in computer science. In fact, we should speak rather of 'theories' as there are more concepts of complexity stressing its different aspects. There are some connections between our problem and questions raised in the theory of *algorithmic* complexity. Kolmogorov, Chaitin and Solomonoff suggested in 1960s to define the complexity of a string of data by the length of the shortest binary program that computes the string—see Li and Vitanyi (1993) for a thorough, yet very readable introduction into the field. The related concept of the minimum description length was developed later by Rissanen in a surprisingly universal principle of inference—see Rissannen (1989). On the other hand, the theory of *computational* complexity analyses directly time and space requirements of various classes of algorithms. To show how the theories of algorithmic and computational complexity are related to our problem is, however, beyond the scope of this book.

1.4 Common Approaches to Approximation

The problem of parameter estimation for non-linear, non-Gaussian or high-dimensional models has been pursued in the last three decades intensively and from quite different viewpoints. We outline briefly the major approaches used.

Nonlinear Estimation

In the late 1960s and early 1970s, much effort was spent in engineering science to find a recursive solution to nonlinear identification. To some extent, the research was driven by the practical need of the aerospace industry which looked for reliable algorithms of guidance and navigation. On the theoretical side, the interest was largely stimulated by the success of Kalman filter (Kalman, 1960) in solving linear problems. A variety of methods have been developed then, usually applying one of the follow-

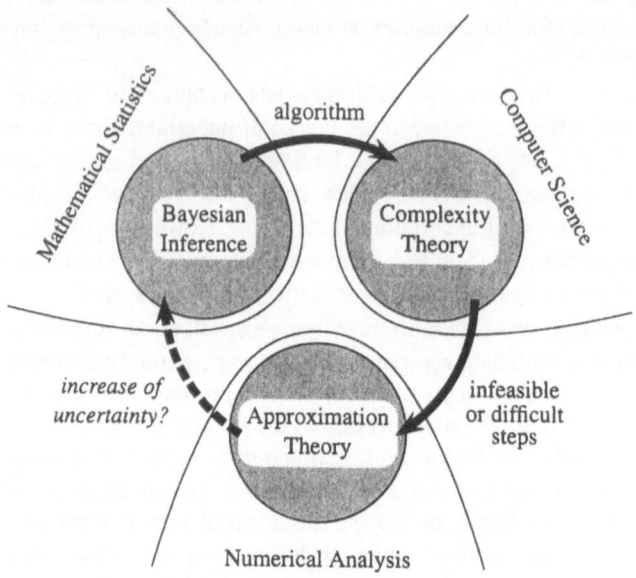

Fig. 1.3. The interdisciplinary character of approximation of Bayesian estimation.

ing two approaches. A detailed survey can be found in Sorenson (1974) and Sorenson (1988).

Direct Approximation of Model. The idea is to substitute a simpler, feasibly identifiable model for the original one. Such a model cannot provide a globally valid description of the identified system and needs to be adapted to changing operating conditions.

Local linearization of the model around the current 'working point' is often the first option to try. When applied to a state-space model, it results in the popular extended Kalman filter, which was treated in detail by, e.g., Sorenson (1966), Jazwinski (1970) and Anderson and Moore (1979).

Other, more sophisticated techniques have been designed. Sorenson and Stubberud (1968) and Athans et al. (1968) used a second-order expansion of nonlinearities. Sorenson and Stubberud (1968) and Srinivasan (1970) employed Gram-Charlier and Edgeworth expansions of a non-Gaussian density.

Recently, the success of neural networks as an effective description of complex mappings has inspired many researchers to study their use in nonlinear modelling and identification—see Chen et al. (1990), Narendra and Parthasarathy (1990), Sanner and Slotine (1991), Billings et al. (1992).

Juditsky et al. (1985) and Sjöberg et al. (1985) give a general account of nonlinear black-box modelling in system identification with particular emphasis on trade-off between the descriptive power and parsimony of the model used.

Functional Approximation of Posterior Density. In contrast to the previous approach, the idea is to keep the original model and to approximate the posterior density instead. With the model untouched, the approach promises to give results closer to the ideal inference although usually at the cost of more computations. Of course, there are many ways of functional approximation that can be applied here, but only some give a good trade-off between accuracy and complexity.

The classical examples of such methods are discrete or point-mass approximation introduced by Bucy (1969) and Bucy and Senne (1971), approximation using mixtures of Gaussian densities developed by Sorenson and Alspach (1971) and Alspach (1974), spline approximation first suggested by de Figueiredo and Jan (1971), generalized least squares approximation shown by Center (1971).

Later, other methods have been published, often as more sophisticated versions of the earlier ones—see, e.g., Wang and Klein (1976), Wiberg (1987), Kramer and Sorenson (1988), Sorenson (1988), Runggaldier and Visentin (1990), Wiberg and DeWolf (1991).

Although some of the methods developed have been remarkably successful in certain applications, many questions still wait for more investigation. In particular, the sequential character of estimation has been given little attention; little is known about possible accumulation of approximation errors. We need to understand better how data compression and numerical errors affect the degree of uncertainty of inference. The lack of theoretical insight is commonly substituted by simulation or practical experience with a particular algorithm.

Practical Bayesian Statistics

Research in Bayesian statistics has focused on the non-recursive case when a complete sample of data is available for analysis. In this case the key problem is effective numerical solution of integrals appearing in calculation of the normalizing constant, marginal density or expectation. Several approaches have been pursued.

Numerical Integration. The complexity of numerical integration depends dramatically on the dimension of the underlying space. Smith et al. (1987) and Naylor and Smith (1988) examined thoroughly several methods of multivariate numerical integration including the product rule strategy based on a Cartesian coordinate system and the spherical rule strategy using a spherical polar coordinate system. In spite of taking additional sophisticated measures like adaptive centering, scaling and orthogonalization, their simulation experiments have indicated hard bounds on the admissible dimension. According to these studies, when functions behave well (possibly after suitable reparametrization), product rules can be effective up to six dimensions and spherical rules up to nine dimensions.

Analytic Approximation. There are many ways in which analytic approximation can help to simplify Bayesian computations. The most obvious but crudest one is normal approximation that is usually substantiated by the (actual or assumed) asymptotic normality of the posterior density. Lindley (1980) developed specific expansions for ratios of integrals appearing commonly in Bayesian estimation. Unfortunately, his approximation requires calculations of up to third-order derivatives which may often be intractable. Tierney and Kadane (1986), Kass et al. (1988), Tierney et al. (1989a), Tierney et al. (1989b) proposed another approach that uses the Laplace method for integrals and requires calculation of only first- and second-order derivatives of (slightly modified) likelihood functions. Their method is probably the most systematically studied and validated analytic approximation today.

Generally speaking, the analytic methods are not so sensitive to the dimension of the problem as the numerical integration procedures, but they can fail when applied to very complex likelihood functions.

Monte Carlo Methods. When the underlying spaces are of high dimension or the functions behave badly, Monte Carlo methods turn out more effective. While the idea of Monte Carlo integration is rather old, recently it has attracted a new interest of the Bayesian community as effective methods of sampling from highly multivariate distributions have appeared.

A typical representative of these methods is the algorithm of Gibbs sampler. Rather than sampling directly from a given distribution, the Gibbs sampler simulates a Markov chain designed so that its stationary distribution equals the required distribution. The Gibbs sampler was introduced formally by Geman and Geman (1984), but the idea dates back to Metropolis et al. (1953). There are similar sampling-based procedures like the substitution or data-augmentation algorithm suggested by Tanner and Wong (1987). The potential of these methods in Bayesian estimation was demonstrated by Gelfand and Smith (1990), Gelfand and Smith (1991), Smith and Gelfand (1992). Recently Carlin et al. (1992) and Gordon et al. (1993) applied sampling-based methods even to recursive estimation.

Note that Geman and Hwang (1986) used in optimization another sampling-based approach that generates samples by simulating a specific stochastic differential equation. This was used elegantly by O'Sullivan et al. (1993).

The sampling-based methods, possibly combined with analytic approximation or numerical integration, are the most effective and most frequently used tools in practical Bayesian statistics today. Gibbs sampler, in particular, has been implemented successfully for extremely high-dimensional problems. Its use is conceptually simple compared to, e.g., advanced numerical integration strategies. It is only a question of time when these methods spread into other areas. Yet, in spite of all the progress made, the methods seem to remain "brute-force" techniques that provide little information to the user about how far the results are from the ideal ones. The usual argument is that with enough samples the approximation can be made arbitrarily accurate, but this can rarely be achieved in real applications.

Nonlinear Filtering

Continuous-time nonlinear filtering is a discipline closely related to discrete-time nonlinear estimation. The continuous-time setting calls, however, for more advanced mathematical tools which makes the field less accessible to outsiders. The subject of nonlinear filtering is analysis and solution of specific stochastic differential equations that govern the evolution of the conditional density. The reader interested in a concise introduction is referred to Davis and Marcus (1981) or Heunis (1990).

As the stochastic differential equations appearing in the theory are typically infinitely-dimensional, reduction of dimensionality is a serious problem. To gain more insight, Brockett (1979; 1980; 1981) and Brockett and Clark (1980) applied in non-linear filtering tools from nonlinear systems theory and theory of Lie algebras. The approach brought a number of penetrating results concerning the existence of exact finite-dimensional filters for certain classes on nonlinear problems (Hazewinkel and Marcus, 1981; Ocone, 1981; Hazewinkel and Marcus, 1982). This work was largely stimulated by the seminal paper by Beneš (1981); another approach to this problem was used by Daum (1988). Although design of approximate finite-dimensional filters has also attracted some attention (Striebel, 1982; Masi and Taylor, 1991), there does not seem to be any universally accepted method yet. A promising concept in this respect is the projection filter proposed by Hanzon (1987) and elaborated in Hanzon and Hut (1991), Brigo (1995) and Brigo et al. (1995).

With respect to our topic, the most notable feature of nonlinear filtering theory is the structural insight gained. The way that differential-geometric and Lie-algebraic methods have been applied in the theory is inspiring even for dicrete-time nonlinear estimation. On the other hand, it is probably fair to say that the nonlinear filtering theory has not bothered much about statistical connections and implications of its results. Most approximate filtering schemes are based on purely algebraic, geometric or analytic arguments and treat approximation separately from inference.

Point Estimation

Classical (frequentist) estimation focuses on design of estimators, i.e., functions that map the observed sample into the parameter space. The estimator approximates the value of the parameter so to optimize explicitly some criterion of goodness of fit. When an explicit model of stochastic components is assumed, the criterion follows from maximum likelihood or maximum a posteriori estimation (Lehmann, 1983; Pázman, 1993). When there is no such model, the choice of the criterion is led pragmatically by the aim to make prediction or equation errors small. Typical representatives are non-linear least squares, prediction error methods and stochastic approximation (Nevel'son and Has'minskiĭ, 1973; Ljung, 1987; Gallant, 1987; Peeters, 1994). Robust statistical methods like M-estimators work with special criteria that are less sensitive to gross errors—outliers (Huber, 1981; Hampel et al., 1986).

Typically, the distribution of the parameter estimator is approximated only asymptotically or not calculated at all. This makes point estimation much simpler to realize. On the other hand, considerably more is known about the asymptotic properties of nonlinear point estimators compared with the approximate Bayesian estimation. Connections between the approximate point estimation and approximate Bayesian estimation seem little understood yet.

Nonlinear Systems and Control Theory

Discrete-time nonlinear estimation has some structural features common with nonlinear systems analysis and control. In the past two decades tremendous progress has been made in understanding the mathematical structure of problems solved in these disciplines—see Jacobs (1980), Byrnes and Lindquist (1986), Byrnes et al. (1988), Isidori (1989), Nijmeijer and van der Schaft (1990). The progress in the area has been propelled by mastering powerful tools from differential geometry, theory of Lie groups and Lie algebras, differential algebra and other fields that traditionally belonged to "pure" mathematics. These advanced tools are likely to have impact on analysis and design of nonlinear estimators as well.

1.5 Why the Geometric Approach?

We have seen that the finiteness of computer resources often forces us to sacrifice the theoretically optimal solution. The typical example is the use of a limited, finite-dimensional statistic in recursive estimation. When the statistic is not sufficient for the model class considered, approximation of the ideal solution is inevitable.

There seems to be little space for approximation in probability-based estimation. Any approximation of the likelihood function or posterior density means violation of the underlying rules of probability theory. Since probability-based estimation does not count essentially with any deviation from the ideal, it gives little idea about consequences of approximation.

Estimation Is Approximation. The bottom line of our approach is the view of estimation as an explicit approximation problem. This view looks natural for estimation schemes that directly match data in Euclidean space, but it is not so obvious when we consider estimation in terms of the likelihood function and posterior density. It can be shown, however, that even the likelihood function and the posterior density are nothing but transformed expressions of a certain "distance" between the empirical and theoretical, model-based probability distributions of data.

This view provides a good start for design of approximate schemes capable of coping with less ideal situations. In particular, when data are compressed and the empirical density is only known to belong to a certain set, it is natural to build approximate estimation upon the minimum "distance" between the model density and the set containing the true empirical density.

When the data statistic is sufficient for the family of model distributions, this approach gives, with a proper "distance", results that coincide with the ideal probabilistic inference. The use of a non-sufficient statistic means necessarily approximation, but approximation with many appealing features.

Background. The approach indicated above dates back to Kullback (1959) and builds upon specific properties of information measures suitable for this purpose, namely Kullback-Leibler distance (Kullback and Leibler, 1951). A fundamental tool in this respect is the Pythagorean relationship that relates Kullback-Leibler distances between three probability density functions similarly as we are used to in Euclidean geometry. The Pythagorean relationship was proved in various contexts by Čencov (1972), Csiszár (1975) and Amari (1985). It makes it possible to develop in probability spaces a specific kind of Pythagorean geometry with deep statistical connections. The minimum Kullback-Leibler distance projections in this geometry are directly related to the maximum likelihood and maximum entropy estimates.

There is also a tight connection between the Pythagorean geometry of estimation and the large deviation theory. The latter provides asymptotic estimates for the probability that the data statistic takes on a given value or, equivalently, that the empirical distribution belongs to a given set. This connection indicates favourable asymptotic properties of the Pythagorean-like approximation.

Major Results. The use of information measures in statistical theory has been attracting a lot of attention, mainly in the context of independent and identically distributed random variables (see, e.g., Chap. 12 in Cover and Thomas, 1991) and Markov chains.

In the following chapters we present a possible extension of the Pythagorean geometry to the case of controlled dynamic systems, with both continuous and discrete data considered. The key points can be summarized as follows.

(a) Inaccuracy is chosen as the starting concept rather than Kullback-Leibler distance in order to include continuous data into consideration smoothly. The use of inaccuracy in estimation is illustrated through a series of examples.

(b) A Pythagorean relationship that links inaccuracies and Kullback-Leibler distance is shown to hold for independent observations. It enables us to decompose the inaccuracy into a sum of two terms, one independent of the unknown parameter, the other independent of data given the value of a certain statistic.

(c) An analogous result for controlled dynamic systems is found as well—by double application of the Pythagorean relationship.

(d) Using the Pythagorean view of estimation, the concept of a sufficient data statistic is given an appealing geometric interpretation.

(e) A sufficient data statistic is shown for controlled dynamic systems too. Because of double application of the Pythagorean relationship, the conditions for the existence of a finite-dimensional sufficient statistic turn out much stricter.

(f) Approximation of the likelihood function and the posterior density given a non-sufficient data statistic is proposed in geometric terms. The solutions for the cases of independent observations and controlled dynamic systems have a formally identical structure.

(g) A proper class of non-sufficient data statistics is shown to coincide with the class of necessary statistics.

(h) As an extension of the above results, the Pythagorean approach is used for approximation of the empirical expectation of an arbitrary function of observed data.

2. From Matching Data to Matching Probabilities

The simplest way of fitting model to given data is to match directly the sequence of observed data with another, model-based sequence so to minimize some measure of mismatch between both. What we show in this chapter is that the 'matching' approach can be extended quite naturally from data to probability density functions of data. The idea is to match the empirical density with a theoretical, model-based one so to minimize some measure of mismatch between both. We shall see that all probability-based methods of parameter estimation do this kind of 'probability matching', implicitly at least.

The data-based and probability-based views of estimation are intimately related for simple models, namely those with linear dynamics and Gaussian stochastics. The situation becomes more difficult and more colourful when one deals with non-linear or non-Gaussian models. The view of parameter estimation via 'probability matching' turns out then more powerful, providing more insight and more guidance. A number of examples in this chapter illustrate the point.

Our primary motivation for the change of viewpoint is, however, to transpose estimation into the form of an explicit approximation problem. This view will serve us later as an ideal start for design of approximations of the classical paradigms.

2.1 Data Matching in Euclidean Space

The well-known least-squares method is used here as a classical example of direct matching of data. The use of geometry in Euclidean space provides a lot of insight.

Uncorrelated Data. The sequence of raw observed data $y^N = (y_1, \ldots, y_N)$, $y_k \in \mathbb{R}$ is usually a too complex object to deal with in subsequent decision-making. To achieve substantial compression of data, we usually fit to y^N another sequence that captures only the major pattern in observed data. The simplest possible model or explanation of data is when all the observed values are matched with a constant

$$y_k = \theta + e_k, \quad k = 1, \ldots, N. \tag{2.1}$$

The variable e_k stands for the error—deviation between the true value y_k and the model-based value θ. The approximation errors are the necessary price for a simpler description of data. The obvious objective is to make the errors as small as possible.

Thinking of the sequence $y^N = (y_1, \ldots, y_N)$ as a (row) vector in $\mathbb{R}^N$, we can measure the total approximation error between y^N and

$$\theta 1^N = \theta(1, 1, \ldots, 1) = \underbrace{(\theta, \theta, \ldots, \theta)}_{N \text{ times}}$$

by the Euclidean norm of a difference of the two vectors

$$\|e^N\|^2 = \|y^N - \theta 1^N\|^2 = \sum_{k=1}^{N} (y_k - \theta)^2.$$

The value of θ that minimizes the total error

$$\hat{\theta}_N = \arg \min_{\theta} \|y^N - \theta 1^N\|^2$$

is the well-known least-squares estimate.

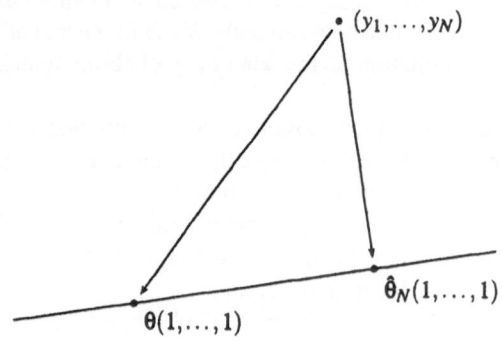

Fig. 2.1. Matching data with constant in Euclidean space.

Linear Regression. The simplest model of data dependence is perhaps the linear fit

$$y_k = \theta y_{k-1} + e_k, \quad k = 2, \ldots, N+1. \tag{2.2}$$

Regarding the sequences y_2^{N+1} and y_1^N as (row) vectors in $\mathbb{R}^N$, we can measure the total approximation error between y_2^{N+1} and y_1^N by the Euclidean norm of their difference

$$\|y_2^{N+1} - \theta y_1^N\|^2 = \sum_{k=2}^{N+1} (y_k - \theta y_{k-1})^2.$$

The value of θ that minimizes the total error

$$\hat{\theta}_N = \arg \min_{\theta} \|y_2^{N+1} - \theta y_1^N\|^2$$

is the least-squares estimate again.

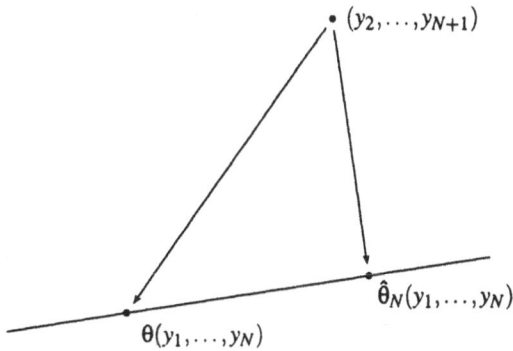

Fig. 2.2. Matching data with time-lagged values in Euclidean space.

General Linear Regression. The above examples can be regarded as special instances of the following model

$$y_{m+1}^{N+m} = o + \sum_{i=1}^{n} \theta_i h_i + e_{m+1}^{N+m} \tag{2.3}$$

where o and h_i, $i = 1, \ldots, n$ are fixed (row) vectors in $\mathbb{R}^N$. The model thus approximates y_{m+1}^{N+m} with a point from a certain subset of $\mathbb{R}^N$ any point of which can be expressed as a sum of a certain fixed point o ("origin" of the subset) and a vector from the vector subspace of $\mathbb{R}^N$ spanned by the vectors h_i, $i = 1, \ldots, n$. Provided the vectors h_i, $i = 1, \ldots, n$ are linearly independent, the subset defined in the above way forms an n-dimensional hyperplane in the Euclidean space $\mathbb{R}^N$. Note the role of $o \neq (0, 0, \ldots, 0)$; it moves the hyperplane into a general position, not necessarily going through the origin of $\mathbb{R}^N$. Thus, o makes the approximating subspace affine.

Note that similarly as in the preceding cases, the value of θ that minimizes the total error

$$\hat{\theta}_N = \arg \min_{\theta} \left\| y_{m+1}^{N+m} - o - \sum_{i=1}^{n} \theta_i h_i \right\|^2$$

is the usual least-squares estimate.

The model (2.3) covers a lot of cases of practical interest, for instance:

- $m = 0$, $n = 1$, $h_1 = (1, 1, \ldots, 1)$
  constant;

- $m = 1$, $n = 1$, $h_1 = (y_1, y_2, \ldots, y_N)$
  first-order autoregression;

- $m = 0$, $n = 1$, $h_1 = (1, 2, \ldots, N)$
  linear trend;

- $m = 2$, $n = 2$, $h_1 = (y_1, y_2, \ldots, y_N)$, $h_2 = (y_2, y_3, \ldots, y_{N+1})$
  second-order autoregression;

- $m = 1, n = 2, h_1 = (y_1, y_2, \ldots, y_N), h_2 = (u_2, u_3, \ldots, u_{N+1})$
  first-order autoregression with external input;

- $m = 1, n = 1, h_1 = (f(y_1), f(y_2), \ldots, f(y_N))$
  first-order nonlinear autoregression given a nonlinear transformation $f(\cdot)$;

- $m = 1, n = 2, h_1 = (f(y_1), f(y_2), \ldots, f(y_N)), h_2 = (g(y_1), g(y_2), \ldots, g(y_N))$
  first-order nonlinear autoregression given nonlinear transformations $f(\cdot)$ and $g(\cdot)$.

The first two cases coincide with (2.1) and (2.2), respectively. Note that the basis vectors h_i may be highly nonlinear in the observed data.

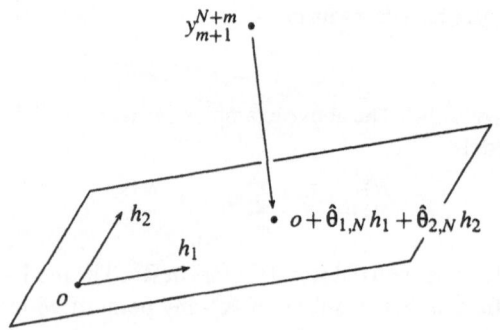

Fig. 2.3. General idea of matching data in Euclidean space.

The model (2.3) can be rewritten pointwise in the following, more common form

$$y_k = \theta^T z_k + e_k, \quad k = m+1, \ldots, N+m \tag{2.4}$$

where the vector z_k is a fixed function of past data. By comparison of (2.3) and (2.4) we easily find the correspondence $z_{k,i} = h_{i,k}$. Thus, for instance, for the third-order linear autoregression we have

$$\begin{bmatrix} y_4, & y_5, & \ldots, & y_{N+3} \end{bmatrix} = \begin{bmatrix} \theta_1, & \theta_2, & \theta_3 \end{bmatrix} \cdot \begin{bmatrix} y_3, & y_4, & \ldots, & y_{N+2} \\ y_2, & y_3, & \ldots, & y_{N+1} \\ y_1, & y_2, & \ldots, & y_N \end{bmatrix} + \begin{bmatrix} e_4, & e_5, & \ldots, & e_{N+3} \end{bmatrix}$$

The rows of the matrix in the above equation correspond to the basis vectors h_3, h_2, h_1

$$h_i = \begin{bmatrix} y_i, y_{i+1}, \ldots, y_{N+i-1} \end{bmatrix}.$$

while its columns stand for $z_{m+1}, z_{m+2}, \ldots, z_{N+m}$, i.e.,

$$z_i = \begin{bmatrix} y_i \\ y_{i-1} \\ y_{i-2} \end{bmatrix}, \quad i = m+1, \ldots, N+m.$$

Nonlinear Regression. The next step in generalizing the idea of matching data in Euclidean space could be to approximate y_{m+1}^{N+m} with a point coming from a more general kind of surface than hyperplane. Such a model may look like

$$y_k = f(\theta, z_k) + e_k, \quad k = m+1, \ldots, N+m. \tag{2.5}$$

Here comes, for instance, nonparametric regression (projection pursuit, multiple adaptive regression splines, hinging hyperplanes) or neural networks.

2.2 Probability-based Estimation

Before we show how 'data matching' can be extended to 'probability matching', we give a brief summary of probability-based estimation. In particular, we derive the likelihood and posterior density for two cases that will be of primary interest to us, namely the case of independent observations and the case of controlled dynamic systems.

Independent Observations

The model situation studied in statistics most intensively is the following. One observes outcomes of a certain random experiment that is repeated in perfectly identical conditions many (possibly infinitely many) times. The outcomes are supposed to be samples from a fixed but unknown probability distribution. The objective is to find the probability distribution.

The assumption of independence of the outcomes in separate trials greatly simplifies analysis of the problem. We use the case throughout the book as a motivational example for the more difficult case of dependent data.

Sample of Data. A *sample* of data is regarded as a sequence of random variables

$$Y^N = (Y_1, \ldots, Y_N)$$

with values in a subset $\mathcal{Y}$ of $\mathbb{R}^{\dim y}$. A sequence of observed (measured) values y_k of $Y_k, k = 1, 2, \ldots, N$

$$y^N = (y_1, \ldots, y_N)$$

is said to be a *realization* of the sample Y^N, an *observed sample* or a *given sample*.

The random variables Y_k are considered *continuous* if not said explicitly else.

Independent and Identically Distributed Data. We assume that Y_k is independent of past data Y^{k-1}

$$Y_k \perp Y^{k-1}, \quad k = 2, \ldots, N. \tag{2.6}$$

In terms of density functions, this reads by (A.27)

$$s_k(y_k | y^{k-1}) = s_k(y_k), \quad k = 2, \ldots, N. \tag{2.7}$$

In addition, we assume that the distribution of Y_k is identical for all k

$$s_k(y) = s(y).$$

Model Family. We assume that the unknown density $s(y)$ comes from a given family

$$\mathcal{S} = \{s_\theta(y) : \theta \in \mathcal{T}\} \tag{2.8}$$

parametrized by the parameter θ taking values in a subset $\mathcal{T}$ of $\mathbb{R}^{\dim \theta}$.

For simplicity, we assume that

$$s_\theta(y) > 0$$

for all $y \in \mathcal{Y}$ and $\theta \in \mathcal{T}$.

The objective of parameter estimation is to guess on a proper value of the parameter θ given an observed sample y^N.

Joint Density. By chain rule (A.24) and independence assumption (2.7), the joint density $q_\theta^N(y^N)$ of Y^N can be expressed as

$$
\begin{aligned}
q_\theta^N(y^N) &= s_\theta(y_1)\, s_\theta(y_2|y_1)\, s_\theta(y_3|y_2,y_1) \ldots s_\theta(y_N|y_{N-1},\ldots,y_1) \\
&= s_\theta(y_1)\, s_\theta(y_2)\, s_\theta(y_3) \ldots s_\theta(y_N) \\
&= \prod_{k=1}^{N} s_\theta(y_k).
\end{aligned}
$$

Hence, we have

$$\boxed{q_\theta^N(y_1,\ldots,y_N) = \prod_{k=1}^{N} s_\theta(y_k).} \tag{2.9}$$

Likelihood Function. When the joint density $q_\theta^N(y^N)$ is regarded as a function of θ for given y^N, it is called a *likelihood function* or simply *likelihood*

$$l_N(\theta) \triangleq q_\theta^N(y^N).$$

It is worth stressing that even if the unknown parameter θ is interpreted as a random variable Θ, likelihood is *not* a probability; the value

$$\int_A l_N(\theta)\, d\theta$$

does not say anything about the probability that Θ belongs to a ceratin set $A \subset \mathcal{T}$. Yet, owing to its definition, likelihood makes it possible to compare different models $s_\theta(y)$, $s_{\theta'}(y)$ relatively to each other; the likelihood ratio

$$\frac{l_N(\theta)}{l_N(\theta')} = \frac{q_\theta^N(y_1,\ldots,y_N)}{q_{\theta'}^N(y_1,\ldots,y_N)} > 1$$

for some θ, $\theta' \in \mathcal{T}$ leads us to prefer $s_\theta(y)$ to $s_{\theta'}(y)$ unless we have enough prior information against this conclusion.

Example 2.1 (*Normal distribution*) Consider the model

$$Y_k = \theta + E_k, \quad E_k \sim N(0, \sigma^2)$$

where $E_1, E_2, \ldots$ are independent, normally distributed random variables with zero mean and known variance σ^2 with a common density function

$$p(e) = \frac{1}{\sqrt{2\pi\sigma^2}} \exp\left(-\frac{1}{2\sigma^2} e^2\right).$$

Since the transformation $e = y - \theta$ has Jacobian equal to one,

$$J(y) = \frac{\partial e}{\partial y} = 1,$$

the density function of the transformed variable $Y_k = \theta + E_k$ is

$$s_\theta(y) = \frac{1}{\sqrt{2\pi\sigma^2}} \exp\left(-\frac{1}{2\sigma^2}(y - \theta)^2\right).$$

The likelihood for a sample y^N is

$$
\begin{aligned}
l_N(\theta) &= \prod_{k=1}^{N} \frac{1}{\sqrt{2\pi\sigma^2}} \exp\left(-\frac{1}{2\sigma^2}(y_k - \theta)^2\right) \\
&= \left(\frac{1}{\sqrt{2\pi\sigma^2}}\right)^N \exp\left(-\frac{1}{2\sigma^2} N \sum_{k=1}^{N}(y_k - \theta)^2\right) \\
&= \left(\frac{1}{\sqrt{2\pi\sigma^2}}\right)^N \exp\left(-\frac{1}{2\sigma^2} N V_N\right) \exp\left(-\frac{1}{2\sigma^2} N(\bar{y}_N - \theta_i)^2\right)
\end{aligned}
$$

where

$$
\begin{aligned}
\bar{y}_N &= E_N(Y) = \frac{1}{N} \sum_{k=1}^{N} y_k \\
V_N &= E_N(Y^2) - E_N(Y)^2
\end{aligned}
$$

stand for the first two empirical moments of Y. □

Posterior Density. When the unknown parameter θ is interpreted as a random variable Θ, it is natural to describe the available information about Θ by its *posterior* density conditional on the observed sample y^N

$$p_N(\theta) \stackrel{\Delta}{=} p(\theta|y^N).$$

Given a certain prior density

$$p_0(\theta) \stackrel{\Delta}{=} p(\theta),$$

the posterior density can be found by applying Bayes's theorem (A.26) and substituting for the joint density of sample $q_\theta^N(y^N)$ from (2.9)

$$p_N(\theta) \propto p_0(\theta)\, q_\theta^N(y^N)$$

$$\propto p_0(\theta)\, l_N(\theta)$$

$$\propto p_0(\theta) \prod_{k=1}^{N} s_\theta(y_k).$$

The symbol $\propto$ stands for equality up to a normalizing factor. As a result, we have

$$p_N(\theta) \propto p_0(\theta) \prod_{k=1}^{N} s_\theta(y_k). \tag{2.10}$$

A recursive formula for the posterior density reads

$$p_k(\theta) \propto p_{k-1}(\theta)\, s_\theta(y_k)$$

for $k = 1, \ldots, N$.

Controlled Dynamic Systems

The basic problem of system identification is to fit a proper model to a dynamic, possibly controlled system. The models used in system identification typically describe the dependence of the system 'output' on its past values and possibly on some external 'inputs' as well. The "conditional" character of the model and the existence of external inputs, measured but typically unmodelled, makes the problem of parameter estimation in general more difficult compared with the independent observations. The difference will be elucidated in detail in Chap. 3 and 4.

Sample of Data. Consider a system on which two sequences of random variables are measured

$$Y^{N+m} = (Y_1, \ldots, Y_{N+m}), \quad U^{N+m} = (U_1, \ldots, U_{N+m})$$

which take values in subsets $\mathcal{Y}$ and $\mathcal{U}$ of $\mathbb{R}^{\dim y}$ and $\mathbb{R}^{\dim u}$, respectively. U_k is defined as a directly manipulated input to the system at time k while Y_k is the output—response of the system at time k to the past history of data represented by the sequences Y^{k-1} and U^k. Both the above sequences form together a complete *sample* of data.

A sequence of observed (measured) values

$$y^{N+m} = (y_1, \ldots, y_{N+m}), \quad u^{N+m} = (u_1, \ldots, u_{N+m})$$

is called a *realization* of the sample Y^{N+m}, U^{N+m}, an *observed sample* or a *given sample*.

The random variables Y_k, U_k, Z_k are considered *continuous* if not said otherwise.

Regression-type Dependence. Suppose that the output values Y_k depend on a limited amount of past data, namely Y_{k-m}^{k-1}, U_{k-m}^k through a known vector function $Z_k = z(U^k, Y^{k-1})$ taking values in a subset $\mathcal{Z}$ of $\mathbb{R}^{\dim z}$. More precisely, assume that Y^k is conditionally independent of Y^{k-1}, U^k given $Z_k = z_k$

$$Y^k \perp Y^{k-1}, U^k \mid Z_k, \quad k = m+1, \ldots, N+m. \tag{2.11}$$

In terms of density functions, it reads by (A.28)

$$s_k(y_k \mid y^{k-1}, u^k) = s_k(y_k \mid z_k), \quad k = m+1, \ldots, N+m. \tag{2.12}$$

In addition, we assume that the conditional distribution of Y_k given $Z_k = z_k$ is identical for all k

$$s_k(y \mid z) = s(y \mid z).$$

Finally, it is assumed that (y_N, z_N) is recursively computable given its last value (y_{N-1}, z_{N-1}) and the latest data (y_N, u_N), i.e., there exists a map F such that

$$(y_N, z_N) = F\big((y_{N-1}, z_{N-1}), (y_N, u_N)\big).$$

Model Family. We suppose that the density $s(y \mid z)$ comes from a given family

$$\mathcal{S} = \{s_\theta(y \mid z) : \theta \in \mathcal{T}\} \tag{2.13}$$

parametrized by the parameter θ taking values in a subset $\mathcal{T}$ of $\mathbb{R}^{\dim \theta}$.

For simplicity, we assume that

$$s_\theta(y \mid z) > 0$$

for all $(y, z) \in \mathcal{Y} \times \mathcal{Z}$ and all $\theta \in \mathcal{T}$.

The objective of parameter estimation is to find a proper value of the parameter θ given observed samples y^{N+m} and u^{N+m}.

Natural Conditions of Control. Let the dependence of the input U_k on the past data Y^{k-1}, U^{k-1} and the parameter θ be expressed through a conditional density $\gamma_k(u_k \mid y^{k-1}, u^{k-1}, \theta)$. In many cases of practical interest, we may adopt a simplifying assumption that the only information about θ used for computation of the new input is the information contained in the past data.

More precisely, we assume that U_k and Θ, interpreted as a random variable, are conditionally independent given $Y^{k-1} = y^{k-1}$, $U^{k-1} = u^{k-1}$

$$U_k \perp \Theta \mid Y^{k-1}, U^{k-1}, \quad k = m+1, \ldots, N+m. \tag{2.14}$$

In terms of density functions, it implies by (A.28)

$$\gamma_k(u_k \mid y^{k-1}, u^{k-1}, \theta) = \gamma_k(u_k \mid y^{k-1}, u^{k-1}), \quad k = m+1, \ldots, N+m. \tag{2.15}$$

Note that the condition (2.15) introduced by Peterka (1981) is really natural in control of technological processes. The condition is clearly satisfied when the input is produced by an open-loop input generator, a closed-loop fixed controller (pretuned using prior information) or closed-loop adaptive controller (based on prior information *and* observed data).

An example when the condition (2.15) is violated is the behaviour of intelligent subjects on the stock market. It is easy to imagine situations when one can profit by observing the other subjects' behaviour rather than by analysing data themselves. In cases like that even the input generator $\gamma_k(u_k|y^{k-1}, u^{k-1}, \theta)$ needs to be modelled and identified. In the sequel, we consider only the cases when (2.15) is satisfied.

Joint Density. Applying chain rule (A.24), the conditional independence assumption (2.12) and assuming the natural conditions of control (2.15), we can rewrite the joint density of Y_{m+1}^{N+m} and U_{m+1}^{N+m} conditional on m initial values of Y_k and U_k as follows

$$q_\theta^N(y_{m+1}^{N+m}, u_{m+1}^{N+m}|y^m, u^m) = \prod_{k=m+1}^{N+m} s_\theta(y_k|y^{k-1}, u^k) \prod_{k=m+1}^{N+m} \gamma_k(u_k|y^{k-1}, u^{k-1}, \theta)$$

$$= \prod_{k=m+1}^{N+m} s_\theta(y_k|z_k) \prod_{k=m+1}^{N+m} \gamma_k(u_k|y^{k-1}, u^{k-1}).$$

We have thus

$$\boxed{q_\theta^N(y_{m+1}^{N+m}, u_{m+1}^{N+m}|y^m, u^m) = \prod_{k=m+1}^{N+m} s_\theta(y_k|z_k) \prod_{k=m+1}^{N+m} \gamma_k(u_k|y^{k-1}, u^{k-1}).} \qquad (2.16)$$

Likelihood Function. The joint density $q_\theta^N(y_{m+1}^{N+m}, u_{m+1}^{N+m}|y^m, u^m)$ conditional on the initial values y^m and u^m is called a *likelihood function* when regarded as a function of θ for given y^{N+m}, u^{N+m}

$$l_N(\theta) \triangleq q_\theta^N(y_{m+1}^{N+m}, u_{m+1}^{N+m}|y^m, u^m).$$

Note that we use the subscript N to indicate N data points

$$(y_{m+1}, z_{m+1}), \ldots, (y_{N+m}, z_{N+m})$$

the likelihood function is based on.

Once again, we stress that likelihood should not be confused with probability. It may be used, however, to compare different models $s_\theta(y|z)$, $s_{\theta'}(y|z)$ relatively to each other; the likelihood ratio

$$\frac{l_N(\theta)}{l_N(\theta')} = \frac{q_\theta^N(y_{m+1}^{N+m}, u_{m+1}^{N+m}|y^m, u^m)}{q_{\theta'}^N(y_{m+1}^{N+m}, u_{m+1}^{N+m}|y^m, u^m)} > 1$$

suggests to prefer $s_\theta(y|z)$ to $s_{\theta'}(y|z)$ unless prior information speaks against it.

Example 2.2 (*Linear regression with Gaussian noise*) Consider the model

$$Y_k = \theta^T Z_k + E_k, \quad E_k \sim N(0, \sigma^2)$$

where θ and Z_k are column vectors and $E_1, E_2, \ldots$ are independent, normally distributed random variables with zero mean and known variance σ^2 with a common density function

$$p(e) = \frac{1}{\sqrt{2\pi\sigma^2}} \exp\left(-\frac{1}{2\sigma^2} e^2\right).$$

Since the transformation $e = y - \theta^T z$ has Jacobian equal to one,

$$J(y) = \frac{\partial e}{\partial y} = 1,$$

the conditional density function of Y_k given $Z_k = z_k$ is

$$s_\theta(y|z) = \frac{1}{\sqrt{2\pi\sigma^2}} \exp\left(-\frac{1}{2\sigma^2}(y - \theta^T z)^2\right).$$

The likelihood for the sample y^{N+m} is

$$
\begin{aligned}
l_N(\theta) &= \prod_{k=m+1}^{N+m} \frac{1}{\sqrt{2\pi\sigma^2}} \exp\left(-\frac{1}{2\sigma^2}(y_k - \theta^T z_k)^2\right) \\
&= \left(\frac{1}{\sqrt{2\pi\sigma^2}}\right)^N \exp\left(-\frac{1}{2\sigma^2} N \sum_{k=m+1}^{N+m} (y_k - \theta^T z_k)^2\right) \\
&= \left(\frac{1}{\sqrt{2\pi\sigma^2}}\right)^N \exp\left(-\frac{1}{2\sigma^2} N V_N\right) \exp\left(-\frac{1}{2\sigma^2} N (\theta - \hat{\theta}_N)^T C_N (\theta - \hat{\theta}_N)\right)
\end{aligned}
$$

with

$$
\begin{aligned}
\hat{\theta}_N &= C_N^{-1} E_N(ZY) \\
V_N &= E_N(Y^2) - E_N(YZ^T) C_N^{-1} E_N(ZY) \\
C_N &= E_N(ZZ^T)
\end{aligned}
$$

where $E_N(Y) = \frac{1}{N} \sum_{k=m+1}^{N+m} y_k$ stands for the empirical mean of a random variable Y. In the above C_N is assumed positive definite. $\square$

Posterior Density. When the unknown parameter θ is treated as a random variable Θ, it is natural to describe its uncertainty through the *posterior* density conditional on the observed samples y^{N+m} and u^{N+m}

$$p_N(\theta) \triangleq p(\theta|y^{N+m}, u^{N+m}).$$

The subscript N indicates again conditioning on N data points

$$(y_{m+1}, z_{m+1}), \ldots, (y_{N+m}, z_{N+m}).$$

Given a prior density conditional on available prior information *and* possibly m initial values y^m, u^m

$$p_0(\theta) \stackrel{\triangle}{=} p(\theta|y^m, u^m),$$

the posterior density can be found by applying Bayes's theorem (A.26). Substituting for the joint density of sample $q_\theta^N(y_{m+1}^{N+m}, u_{m+1}^{N+m}|y^m, u^m)$ from (2.16) and assuming the natural conditions of control (2.15) satisfied, we have

$$p_N(\theta) \propto p_0(\theta) q_\theta^N(y_{m+1}^{N+m}, u_{m+1}^{N+m}|y^m, u^m)$$

$$\propto p_0(\theta) l_N(\theta)$$

$$\propto p_0(\theta) \prod_{k=m+1}^{N+m} s_\theta(y_k|z_k) \prod_{k=m+1}^{N+m} \gamma_k(u_k|y^{k-1}, u^{k-1}, \theta)$$

$$\propto p_0(\theta) \prod_{k=m+1}^{N+m} s_\theta(y_k|z_k) \prod_{k=m+1}^{N+m} \gamma_k(u_k|y^{k-1}, u^{k-1})$$

where $\propto$ stands for equality up to a normalizing factor. It follows that

$$\boxed{p_N(\theta) \propto p_0(\theta) \prod_{k=m+1}^{N+m} s_\theta(y_k|z_k).} \qquad (2.17)$$

The computation of the posterior density can easily be organized recursively

$$p_k(\theta) \propto p_{k-1}(\theta) s_\theta(y_{k+m}|z_{k+m})$$

for $k = 1, \ldots, N$.

Prior Density. The piece of information contained in the initial data y^m, u^m can be used, in principle, to update the prior, unconditional density $p(\theta)$. The Bayes's theorem (A.26) gives the clue

$$p_0(\theta) \propto p(\theta) q_\theta^0(y^m, u^m). \qquad (2.18)$$

In practice, however, the piece of information carried by y^m, u^m is usually neglected and we set simply $p_0(\theta) = p(\theta)$. This is what is supposed in the sequel.

The following example illustrates the point.

Example 2.3 (*Stable process*) Consider the first-order autoregressive model

$$Y_k = \theta Y_{k-1} + E_k, \quad E_k \sim N(0, \sigma^2)$$

where $E_1, E_2, \ldots$ is a sequence of independent, normally distributed random variables with zero mean and known and constant variance σ^2. The conditional theoretical density thus takes the form

$$s_\theta(y_k|y_{k-1}) = \frac{1}{\sqrt{2\pi\sigma^2}} \exp\left(-\frac{1}{2\sigma^2}(y_k - \theta y_{k-1})^2\right)$$

Suppose that the regression coefficient θ lies within the interval $(-1, 1)$, in other words, the underlying process is supposed to be stable. This fact can be utilized to specify the density function of the first observation Y_1. Assuming that the underlying process reached its stationary behaviour before being first observed, we can set

$$E(Y_1|\theta) = 0,$$

$$E(Y_1^2|\theta) = \frac{\sigma^2}{1-\theta^2}.$$

Given these moments and assuming normality of Y_1, we have

$$q_\theta^0(y_1) = \frac{\sqrt{1-\theta^2}}{\sqrt{2\pi\sigma^2}} \exp\left(-\frac{1}{2\sigma^2}(1-\theta^2)y_1^2\right).$$

Substitution in (2.18) gives

$$p_0(\theta) \propto p(\theta)\sqrt{1-\theta^2} \exp\left(-\frac{1}{2\sigma^2}(1-\theta^2)y_1^2\right).$$

See Peterka (1981, Example 4.2) for more details.   □

2.3 Estimation via Inaccuracy

Borrowing the notion of inaccuracy from information theory, we can transpose probability-based estimation into the form of an explicit approximation problem.

Independent Observations

Empirical Density. Given the observed sample y^N, an *empirical density* of Y is defined as

$$r_N(y) \overset{\triangle}{=} \frac{1}{N}\sum_{k=1}^{N} \delta(y-y_k) \tag{2.19}$$

where $\delta(y)$ is a Dirac function satisfying $\delta(y) = 0$ for $y \neq 0$ and

$$\int_y \delta(y)\,dy = 1.$$

The empirical density can be updated recursively according to the formula

$$r_k(y) = \frac{k-1}{k}r_{k-1}(y) + \frac{1}{k}\delta(y-y_k) \tag{2.20}$$

for $k = 1,\dots,N$.

Theoretical Density. The empirical density represents a raw description of observed data which is not "contaminated" by any model assumption—except the structural assumption about independent and identically distributed data. Yet, in most applications we prefer to fit to $r_N(y)$ a density $s_\theta(y)$ from a certain parametric family. The density $s_\theta(y)$ is called *theoretical* or *model* or *sampling* density.

There are two fundamental reasons for that. First, by substituting $s_\theta(y)$ for $r_N(y)$, we drastically reduce the complexity of computations. While the whole sample y^N is basically needed to construct $r_N(y)$, the parameter value θ is sufficient to identify the theoretical density $s_\theta(y)$ within a given family $\mathcal{S}$. Second, through the choice of the parametric family $\mathcal{S}$, we bring a substantial piece of prior information into play. While the empirical density $r_N(y)$ describes the past data, the theoretical density $s_\theta(y)$ makes it possible to predict the future behaviour of data as well.

Example 2.4 (*Normal distribution*) Figures 2.4 and 2.5 show samples of 100 and 1000 normally-distributed data, respectively, with mean $\theta = -1$ and variance $\sigma^2 = 3$, produced by a random number generator. Suppose we are to decide between the hypotheses $Y \sim N(-1,3)$ and $Y \sim N(1,3)$ given the observed samples. The traditional way is to compare the residuals $y_k - \theta_i$ for the two hypotheses $i = 0, 1$ with $\theta_0 = -1$ and $\theta_1 = +1$. Another possibility is, however, to compare directly the empirical density $r_N(y)$ (envisaged in figures by histograms) with the theoretical densities $s_{\theta_i}(y)$, $i = 0, 1$ by evaluating the corresponding inaccuracies $K(r_N : s_{\theta_i})$. The latter possibility has a certain appeal. □

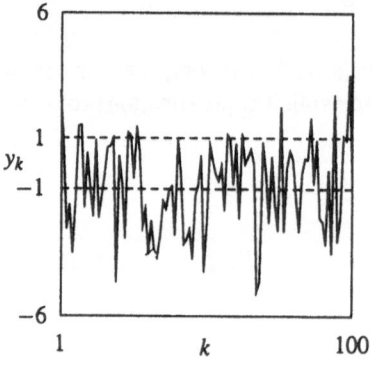 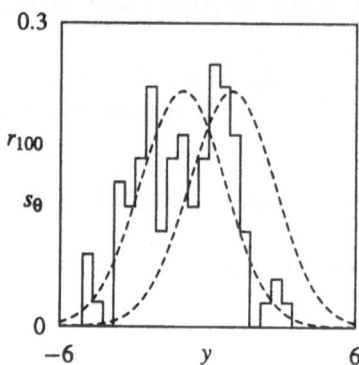

Fig. 2.4. Fitting of normal density to a sample of 100 data generated by normal random number generator.

Inaccuracy. This chapter centres around the notion of *inaccuracy*. We give now its definition and show its role in parameter estimation. In Sect. 2.5 we come back to the notion itself showing its various interpretations.

Inaccuracy of $r_N(y)$ relative to $s_\theta(y)$ is defined as

$$K(r_N : s_\theta) \triangleq \int r_N(y) \log \frac{1}{s_\theta(y)} \, dy. \tag{2.21}$$

Note that throughout the book, the logarithm is always understood to the base e.

Example 2.5 (*Normal distribution*) Given the normal theoretical density

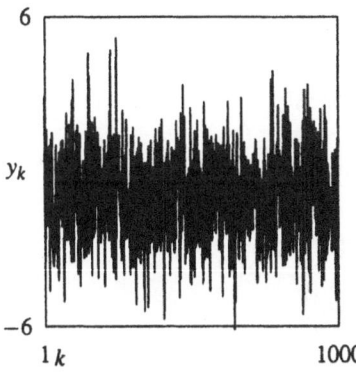 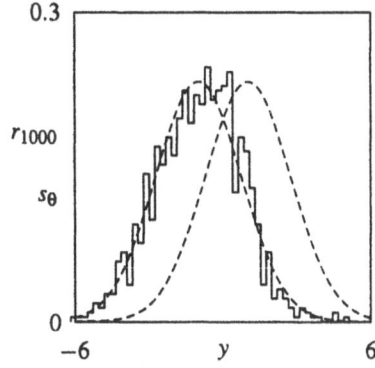

Fig. 2.5. The same problem given a sample of 1000 data.

$$s_\theta(y) = \frac{1}{\sqrt{2\pi\sigma^2}} \exp\left(-\frac{1}{2\sigma^2}(y-\theta)^2\right),$$

we can easily compute

$$
\begin{aligned}
K(r_N:s_\theta) &= -\int r_N(y) \log\left(\frac{1}{\sqrt{2\pi\sigma^2}} \exp\left(-\frac{1}{2\sigma^2}(y-\theta)^2\right)\right) dy \\
&= \frac{1}{2} \log 2\pi\sigma^2 + \int r_N(y) \frac{1}{2\sigma^2}(y-\theta)^2 \, dy \\
&= \frac{1}{2} \log 2\pi\sigma^2 + \frac{1}{2\sigma^2} \frac{1}{N} \sum_{k=1}^{N} (y_k - \theta)^2 \\
&= \frac{1}{2} \log 2\pi\sigma^2 + \frac{1}{2\sigma^2} V_N + \frac{1}{2\sigma^2}(\theta - \bar{y}_N)^2
\end{aligned}
$$

with the empirical moments

$$
\begin{aligned}
\bar{y}_N &= E_N(Y) = \frac{1}{N} \sum_{k=1}^{N} y_k, \\
V_N &= E_N(Y^2) - E_N(Y)^2.
\end{aligned}
$$

With respect to the obvious inequality

$$K(r_N:s_\theta) - K(r_N:s_{\bar{y}_N}) = \frac{1}{2\sigma^2}(\theta - \bar{y}_N)^2 \geq 0,$$

the minimum of the inaccuracy $K(r_N:s_\theta)$ on $\mathcal{T}$ is attained at $\theta = \bar{y}_N$.   □

Joint Density of Sample. Using inaccuracy, the joint density (2.9) can be rewritten as follows

$$q_\theta^N(y^N) = \prod_{k=1}^N s_\theta(y_k)$$

$$= \exp\left(N\frac{1}{N}\sum_{k=1}^N \log s_\theta(y_k)\right)$$

$$= \exp\left(N\int r_N(y)\log s_\theta(y)\,dy\right)$$

$$= \exp\left(-N\int r_N(y)\log\frac{1}{s_\theta(y)}\,dy\right)$$

$$= \exp\bigl(-NK(r_N{:}s_\theta)\bigr).$$

Note that we made use of the assumption $s_\theta(y_k) > 0$.

As a result, we get a surprisingly simple expression of the joint density

$$\boxed{q_\theta^N(y^N) = \exp\bigl(-NK(r_N{:}s_\theta)\bigr).} \tag{2.22}$$

Likelihood. Recall that *likelihood* $l_N(\theta)$ for a given sample y^N is just the joint density $q_\theta^N(y^N)$ of the sample Y^N taken as a function of the unknown parameter θ. Therefore,

$$\boxed{l_N(\theta) = \exp\bigl(-NK(r_N{:}s_\theta)\bigr).} \tag{2.23}$$

The expression stresses the basic ingredients of probability-based estimation:

$$l_N(\theta) = \exp\bigl(-\;\boxed{N}\;K(\;\boxed{r_N}\;{:}\;\boxed{s_\theta}\;)\bigr)$$

amount   empirical   theoretical
of data   density   density

Posterior Density. Applying Bayes's theorem (A.26) and substituting for the joint density of sample $q_\theta^N(y^N)$ from (2.22), we derive the *posterior* density of Θ conditional on $Y^N = y^N$ in the form

$$p_N(\theta) \propto p_0(\theta)\,q_\theta^N(y^N)$$

$$\propto p_0(\theta)\,l_N(\theta)$$

$$\propto p_0(\theta)\,\exp\bigl(-NK(r_N{:}s_\theta)\bigr)$$

where $\propto$ denotes equality up to a normalizing factor.

Hence, we have an alternative formula

$$\boxed{p_N(\theta) \propto p_0(\theta)\,\exp\bigl(-NK(r_N{:}s_\theta)\bigr)} \tag{2.24}$$

which makes the key ingredients of Bayesian estimation clearly separated:

$$p_N(\theta) \propto \boxed{p_0(\theta)}\,\exp\bigl(-\;\boxed{N}\;K(\;\boxed{r_N}\;{:}\;\boxed{s_\theta}\;)\bigr)$$

prior   amount   empirical   theoretical
density   of data   density   density

Conjugate Prior. It is convenient if the prior density $p_0(\theta)$ is chosen from a *conjugate* family—closed under conditioning on observed data (cf. Robert, 1989). The expression (2.24) suggests a general form of conjugate priors

$$p_v(\theta) \propto \exp\left(-v K(r_v : s_\theta)\right) \tag{2.25}$$

where $r_v(y)$ stands for a "prior" density of Y based on v actual or fictitious observations. Thus, (2.25) can be interpreted as a "posterior" density given a uniform prior $p_0(\theta) \propto 1$ and v data with the empirical density $r_v(y)$. To be consistent with this view, we assume that v is nonnegative but not necessarily integer. By choosing a particular value of v we put more or less weight (prior belief) on $r_v(y)$.

The above view is appealing though not quite common. Instead of expressing prior knowledge in terms of a distribution of parameters of a particular model, we suggest to express it directly through a distribution of observed data. In such a way, we can consider different model classes and still keep a single description of prior information. The formula (2.25) shows the prior density of Y can be converted into a prior density of Θ given a particular model.

It is easy to verify that given a conjugate prior (2.25) the posterior density (2.24) preserves the form of the prior

$$
\begin{aligned}
p_{v+N}(\theta) &\propto p_v(\theta) \exp\left(-N K(r_N : s_\theta)\right) \\
&\propto \exp\left(-v K(r_v : s_\theta)\right) \exp\left(-N K(r_N : s_\theta)\right) \\
&\propto \exp\left(-(v + N) K(r_{v+N} : s_\theta)\right)
\end{aligned} \tag{2.26}
$$

where

$$r_{v+N}(y) = \frac{v}{v+N} r_v(y) + \frac{N}{v+N} r_N(y) \tag{2.27}$$

is now a *mixture*, i.e., convex combination of the prior density $r_v(y)$ and the empirical density $r_N(y)$. In accordance with our intuition, the weight on $r_v(y)$ tends to zero as $N \to \infty$.

Note that using the conjugate prior (2.25) we are able to keep the symmetry of data and model which is normally lost in the Bayesian scheme:

$$p_{v+N}(\theta) \propto \exp\left(- \boxed{(v+N)} \, K(\, \boxed{r_{v+N}} : \boxed{s_\theta} \,)\right)$$

amount of data   modified empirical density   theoretical density

Example 2.6 (*Normal distribution*) Consider the normal sampling density

$$s_\theta(y) = \frac{1}{\sqrt{2\pi\sigma^2}} \exp\left(-\frac{1}{2\sigma^2}(y - \theta)^2\right).$$

By Example 2.5, inaccuracy relative to the normal density is

$$K(r_v : s_\theta) = C + \frac{1}{2\sigma^2}(\theta - \bar{y}_v)^2$$

where C is independent of θ. The conjugate prior (2.25) thus takes the form

$$p_{\rm v}(\theta) \propto \exp\left(-\frac{1}{2\sigma^2}\,{\rm v}\,(\theta - \bar{y}_{\rm v})^2\right)$$

which coincides with the usual choice.   □

Controlled Dynamic Systems

Empirical Density. Given the sample y^{N+m}, u^{N+m}, a *joint empirical density* of (Y, Z) is defined as

$$r_N(y, z) = \frac{1}{N}\sum_{k=m+1}^{N+m} \delta(y - y_k, z - z_k) \tag{2.28}$$

where $\delta_{y,z}$ is a Dirac function satisfying $\delta(y, z) = 0$ for $y \neq 0$ or $z \neq 0$ and

$$\iint_{\mathcal{Y}\times\mathcal{Z}} \delta(y, z)\,{\rm d}y\,{\rm d}z = 1.$$

Similarly as with likelihood and posterior density, we use the subscript N to indicate the number of data points

$$(y_{m+1}, z_{m+1}), \ldots, (y_{N+m}, z_{N+m})$$

the empirical density is based on.

The empirical density can be updated recursively according to

$$r_k(y, z) = \frac{k-1}{k}\,r_{k-1}(y, z) + \frac{1}{k}\delta(y - y_{k+m}, z - z_{k+m}) \tag{2.29}$$

for $k = 1, \ldots, N$.

In the sequel, we shall often need the *marginal empirical density* of Z

$$\begin{aligned}
\bar{r}_N(z) &= \int r_N(y, z)\,{\rm d}y \\
&= \frac{1}{N}\sum_{k=m+1}^{N+m}\delta(z - z_k). \tag{2.30}
\end{aligned}$$

Theoretical Density. The joint empirical density $r_N(y, z)$ provides again a raw description of observed data which does not use any model assumption except the structural assumption about the conditional independence, i.e., the definition of Z. For reasons analogous to those said for independent observations, it is normally preferred to fit to $r_N(y, z)$ a *theoretical density* or *model* or *sampling* density $s_\theta(y|z)$. Note that the theoretical density is in this case conditional, modelling only the dependence of Y_k on Z_k.

Example 2.7 (*Linear autoregression with Cauchyian noise*) Consider the first-order autoregressive model

$$Y_k = \theta Y_{k-1} + E_k, \quad E_k \sim C(0,1)$$

where $E_1, E_2, \ldots$ are independent, Cauchy-distributed random variables with a common density function

$$n(e) = \frac{1}{\pi} \frac{1}{1+e^2}.$$

Since the transformation $e = y - \theta z$ has Jacobian equal to one,

$$J(y) = \frac{\partial e}{\partial y} = 1,$$

the conditional density function of the transformed variable Y_k given $Z_k = Y_{k-1}$ is

$$s_\theta(y|z) = \frac{1}{\pi} \frac{1}{1+(y-\theta z)^2}.$$

The assumption of Cauchy distribution of the noise component models the appearance of outliers in observed data. Fig. 2.6 shows a simulated sample y^{101} of 101 data for $\theta = 0.98$. The Cauchyian noise causes abrupt changes of the process. The outliers are easy to recognize in the scatter plot of y_k against y_{k-1}. The scatter plot of (y_k, z_k) gives us a good idea about the empirical density $r_N(y,z)$ as well.   □

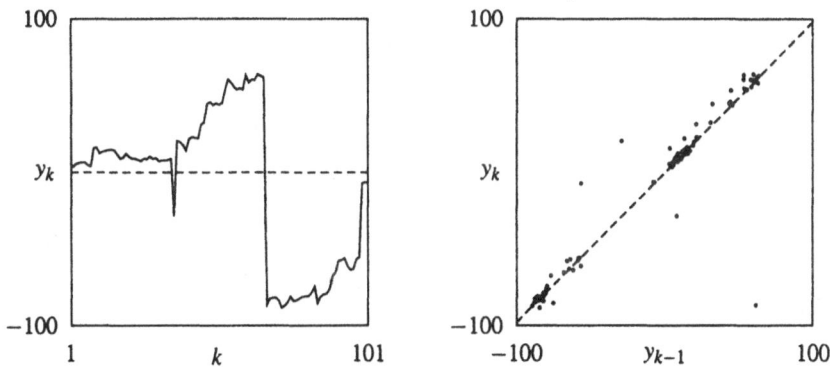

Fig. 2.6. Linear autoregression with Cauchyian noise: a sequence of 101 samples of $Y_k = 0.98 Y_{k-1} + E_k$ with Cauchy-distributed noise $E_k \sim C(0,1)$.

Conditional Inaccuracy. Dealing with the joint empirical density $r_N(y,z)$ and a conditional theoretical density $s_\theta(y|z)$, we need to introduce the notion of *conditional inaccuracy*.

Conditional inaccuracy of $r_N(y,z)$ relative to $s_\theta(y|z)$ is defined as

$$\bar{K}(r_N : s_\theta) \triangleq \iint r_N(y,z) \log \frac{1}{s_\theta(y|z)} \, dy \, dz. \tag{2.31}$$

Example 2.8 (*Linear regression with Gaussian noise*)   Given the conditional theoretical density

$$s_\theta(y|z) = \frac{1}{\sqrt{2\pi\sigma^2}} \exp\left(-\frac{1}{2\sigma^2}(y - \theta^T z)^2\right),$$

the conditional inaccuracy takes the form

$$\bar{K}(r_N:s_\theta) = \frac{1}{2}\log 2\pi\sigma^2 + \frac{1}{2\sigma^2}V_N + \frac{1}{2\sigma^2}(\theta - \hat{\theta}_N)^T C_N (\theta - \hat{\theta}_N)$$

with

$$\hat{\theta}_N = C_N^{-1} E_N(ZY)$$
$$V_N = E_N(Y^2) - E_N(YZ^T) C_N^{-1} E_N(ZY)$$
$$C_N = E_N(ZZ^T)$$

where $E_N(Y) = \frac{1}{N}\sum_{k=m+1}^{N+m} y_k$ stands for the empirical mean of a random variable Y. The matrix C_N is supposed positive definite above.

Owing to the obvious inequality

$$\bar{K}(r_N:s_\theta) - \bar{K}(r_N:s_{\hat{\theta}_N}) = \frac{1}{2\sigma^2}(\theta - \hat{\theta}_N)^T C_N (\theta - \hat{\theta}_N) \geq 0,$$

the minimum of inaccuracy on $\mathcal{T}$ is attained at $\theta = \hat{\theta}_N$. $\square$

Joint Density of Sample.   By (2.16) the joint density of sample is

$$q_\theta^N(y_{m+1}^{N+m}, u_{m+1}^{N+m}|y^m, u^m) = \Gamma(y^{N+m}, u^{N+m}) \prod_{k=m+1}^{N+m} s_\theta(y_k|z_k)$$

where

$$\Gamma(y^{N+m}, u^{N+m}) = \prod_{k=m+1}^{N+m} \gamma_k(u_k|y^{k-1}, u^{k-1})$$

is a factor independent of θ. Using conditional inaccuracy, we can write

$$\prod_{k=m+1}^{N+m} s_\theta(y_k|z_k) = \Gamma(y^{N+m}, u^{N+m}) \exp\left(N\frac{1}{N}\sum_{k=1}^{N}\log s_\theta(y_k|z_k)\right)$$
$$= \Gamma(y^{N+m}, u^{N+m}) \exp\left(N \iint r_N(y,z)\log s_\theta(y|z)\,dy\,dz\right)$$
$$= \Gamma(y^{N+m}, u^{N+m}) \exp\left(-N \iint r_N(y,z)\log \frac{1}{s_\theta(y)}\,dy\,dz\right)$$
$$= \Gamma(y^{N+m}, u^{N+m}) \exp\left(-N\bar{K}(r_N:s_\theta)\right).$$

Note that we made use of the assumption $s_\theta(y_k|z_k) > 0$.

As a result, we have the following expression

$$\boxed{q_\theta^N(y_{m+1}^{N+m}, u_{m+1}^{N+m}|y^m, u^m) = \Gamma(y^{N+m}, u^{N+m}) \exp\left(-N\bar{K}(r_N:s_\theta)\right).}$$ (2.32)

Likelihood Function. The *likelihood function* $l_N(\theta)$ for given samples y^{N+m} and u^{N+m}, i.e., the joint density $q_\theta^N(y_{m+1}^{N+m}, u_{m+1}^{N+m} | y^m, u^m)$ taken as a function of the unknown parameter θ takes after substituting from (2.32) the form

$$l_N(\theta) = \Gamma(y^{N+m}, u^{N+m}) \exp\left(-N\bar{K}(r_N \colon s_\theta)\right). \tag{2.33}$$

The expression separates well the basic elements of estimation:

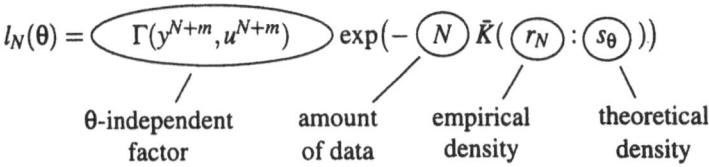

$$l_N(\theta) = \underbrace{\Gamma(y^{N+m}, u^{N+m})}_{\substack{\theta\text{-independent} \\ \text{factor}}} \exp\left(-\underbrace{N}_{\substack{\text{amount} \\ \text{of data}}} \bar{K}(\underbrace{r_N}_{\substack{\text{empirical} \\ \text{density}}} \colon \underbrace{s_\theta}_{\substack{\text{theoretical} \\ \text{density}}}))$$

Posterior Density. Applying Bayes's theorem (A.26) and substituting for the joint density of sample from (2.32), we derive the *posterior* density of Θ conditional on the observed sample y^{N+m}, u^{N+m} in the form

$$
\begin{aligned}
p_N(\theta) &\propto p_0(\theta)\, q_\theta^N(y_{m+1}^{N+m}, u_{m+1}^{N+m} | y^m, u^m) \\
&\propto p_0(\theta)\, l_N(\theta) \\
&\propto p_0(\theta)\, \exp\left(-N\bar{K}(r_N \colon s_\theta)\right)
\end{aligned}
$$

where $\propto$ stands for equality up to a normalizing factor.

The resulting expression

$$p_N(\theta) \propto p_0(\theta)\, \exp\left(-N\bar{K}(r_N \colon s_\theta)\right) \tag{2.34}$$

shows clearly the key elements of Bayesian estimation:

$$p_N(\theta) \propto \underbrace{p_0(\theta)}_{\substack{\text{prior} \\ \text{density}}} \exp\left(-\underbrace{N}_{\substack{\text{amount} \\ \text{of data}}} \bar{K}(\underbrace{r_N}_{\substack{\text{empirical} \\ \text{density}}} \colon \underbrace{s_\theta}_{\substack{\text{theoretical} \\ \text{density}}}))$$

Conjugate Prior. Similarly as for the independent observations, the expression (2.34) suggests a general form of conjugate priors

$$p_\nu(\theta) \propto \exp\left(-\nu\bar{K}(r_\nu \colon s_\theta)\right). \tag{2.35}$$

Here $r_\nu(y, z)$ stands for a "prior" density of (Y, Z) built upon prior information and possibly m initial values y^m and u^m. The scalar ν counts the number of actual or fictitious observations $r_\nu(y, z)$ is built on. The density (2.35) can be regarded as a "posterior" density given a uniform prior $p_0(\theta) \propto 1$ and ν data with the empirical density $r_\nu(y, z)$. The scalar ν is supposed to be nonnegative but not necessarily integer. Its practical meaning is to put more or less weight (prior belief) on $r_\nu(y, z)$.

Given a conjugate prior (2.35), the posterior density (2.34) preserves its form

$$
\begin{aligned}
p_{v+N}(\theta) &\propto p_v(\theta) \exp\left(-N\bar{K}(r_N:s_\theta)\right) \\
&\propto \exp\left(-v\bar{K}(r_v:s_\theta)\right) \exp\left(-N\bar{K}(r_N:s_\theta)\right) \\
&\propto \exp\left(-(v+N)\bar{K}(r_{v+N}:s_\theta)\right)
\end{aligned}
\tag{2.36}
$$

where the modified empirical density

$$
r_{v+N}(y,z) = \frac{v}{v+N} r_v(y,z) + \frac{N}{v+N} r_N(y,z)
\tag{2.37}
$$

is a *mixture* of the prior density $r_v(y,z)$ and the empirical density $r_N(y,z)$. In accordance with our intuition, the weight on $r_v(y,z)$ tends to zero as $N \to \infty$.

Note that using a conjugate prior we are able to keep the symmetry of data and model which is lost in the general Bayesian scheme:

$$
p_{v+N}(\theta) \propto \exp\left(-\; \underbrace{(v+N)}\;\; \bar{K}(\; \underbrace{r_{v+N}} : \underbrace{s_\theta}\;)\right)
$$

amount	empirical	theoretical
of data	density	density

Notice the amazing similarity of results for the independent observations and controlled dynamic systems. The only difference is that the latter requires the use of conditional inaccuracy to compare the joint empirical density $r_N(y,z)$ with the conditional theoretical density $s_\theta(y|z)$. We shall see later that the difference is crucial.

2.4 Maximum Likelihood versus Bayesian Estimation

In terms of inaccuracy, two basic approaches to parameter estimation can be characterized as follows.;

(a) We are interested only in finding the theoretical density that minimizes the inaccuracy $K(r_N:s_\theta)$ or the conditional inaccuracy $\bar{K}(r_N:s_\theta)$.

(b) We want to evaluate the inaccuracy $K(r_N:s_\theta)$ or the conditional inaccuracy $\bar{K}(r_N:s_\theta)$ for all possible theoretical densities, i.e., as a function of θ.

The former view can be shown equivalent to maximum likelihood estimation, the latter approach is the core of Bayesian estimation.

Independent Observations

Maximum Likelihood Estimation. The expression (2.23)

$$
l_N(\theta) = \exp\left(-N K(r_N:s_\theta)\right)
$$

implies the following characterization of inaccuracy

$$K(r_N{:}s_\theta) = -\frac{1}{N} \log l_N(\theta).$$

Since logarithm is a monotonous function, the maximum likelihood (ML) estimate

$$\hat{\theta}_{ML} = \arg \max_\theta l_N(\theta)$$

is equivalent to the the minimum inaccuracy estimate

$$\boxed{\hat{\theta}_{ML} = \arg \min_{\theta \in \mathcal{T}} K(r_N{:}s_\theta)} \tag{2.38}$$

provided the extremum point exists.

This observation allows us to see maximum likelihood estimation as minimum inaccuracy approximation (see Fig. 2.7).

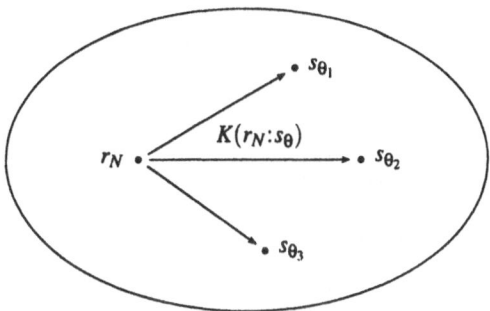

Fig. 2.7. A schematic view of minimum inaccuracy estimation for independent observations. The smaller the inaccuracy $K(r_N{:}s_\theta)$ of $r_N(y)$ relative to $s_\theta(y)$ is, the better is the match between data and model.

Maximum A Posteriori Estimation. A natural Bayesian counterpart of the maximum likelihood estimate is the maximum a posteriori probability (MAP) estimate defined as

$$\hat{\theta}_{MAP} \overset{\triangle}{=} \arg \max_\theta p_N(\theta).$$

The posterior expression (2.24)

$$p_N(\theta) \propto p_0(\theta) \exp\left(-N K(r_N{:}s_\theta)\right)$$

shows that the MAP estimate is equivalent to

$$\hat{\theta}_{MAP} = \arg \min_{\theta \in \mathcal{T}} \left(K(r_N{:}s_\theta) - \frac{1}{N} \log p_0(\theta)\right) \tag{2.39}$$

if the extremum point exists. Compared with the maximum likelihood estimation
(2.38), inaccuracy is modified here by the normalized log-prior. The effect of this
term can be important for short samples; when N gets large enough, the term becomes
typically negligible.

Using the conjugate prior (2.25)

$$p_v(\theta) \propto \exp\left(-\nu K(r_v:s_\theta)\right),$$

we can express the MAP estimate (2.39) in a more compact form

$$\hat{\theta}_{MAP} = \arg \min_{\theta \in \mathcal{T}} K(r_{v+N}:s_\theta). \tag{2.40}$$

Bayesian Estimation. Note that both the ML and MAP estimation schemes are
searching only for a point $s_{\hat{\theta}}(y)$ (not necessarily single one) relative to which inaccuracy of the empirical density $r_N(y)$ or $r_{v+N}(y)$ is minimized.

Compared with the point estimation, Bayesian estimation is by far more ambi-
tious. Its objective is to evaluate the functions

$$K(r_N:s_\theta), \quad \theta \in \mathcal{T}$$

or

$$K(r_{v+N}:s_\theta), \quad \theta \in \mathcal{T},$$

i.e., inaccuracy of $r_N(y)$ or $r_{v+N}(y)$ relative to *all* theoretical densities $s_\theta(y)$, $\theta \in \mathcal{T}$.
The posterior density can be constructed from the above functions easily through the
maps (2.24) and (2.26), respectively.

Controlled Dynamic Systems

Maximum Likelihood Estimation. The expression (2.33) of likelihood

$$l_N(\theta) = \Gamma(y^{N+m}, u^{N+m}) \exp\left(-N\bar{K}(r_N:s_\theta)\right).$$

makes it possible to characterize the conditional inaccuracy as follows

$$\bar{K}(r_N:s_\theta) = -\frac{1}{N} \log l_N(\theta) + \frac{1}{N} \log \Gamma(y^{N+m}, u^{N+m}).$$

Since logarithm is monotonous and the factor $\Gamma(\cdot)$ is independent of θ, the maximum
likelihood (ML) estimate

$$\hat{\theta}_{ML} = \arg \max_\theta l_N(\theta)$$

is equivalent to the minimum *conditional* inaccuracy estimate

$$\hat{\theta}_{ML} = \arg \min_{\theta \in \mathcal{T}} \bar{K}(r_N:s_\theta) \tag{2.41}$$

provided the extremum point exists.

It is worth noting that given any marginal "weighting" density $w(z)$ it holds

$$
\begin{aligned}
K(r_N{:}s_\theta w) &= \iint r_N(y,z) \log \frac{1}{s_\theta(y|z)\, w(z)} \, dy\, dz \\
&= \iint r_N(y,z) \log \frac{1}{s_\theta(y|z)} \, dy\, dz + \int \tilde{r}_N(z) \log \frac{1}{w(z)} \, dz \\
&= \bar{K}(r_N{:}s_\theta) + K(\tilde{r}_N{:}w).
\end{aligned}
$$

Hence,

$$
K(r_N{:}s_\theta w) = C + \bar{K}(r_N{:}s_\theta)
$$

where C is a constant independent of θ. The estimate minimizing $\bar{K}(r_N{:}s_\theta)$ is thus equivalent to the estimate minimizing $K(r_N{:}s_\theta w)$ for any fixed $w(z)$. It allows us to view maximum likelihood estimation as minimum conditional inaccuracy approximation (see Fig. 2.8).

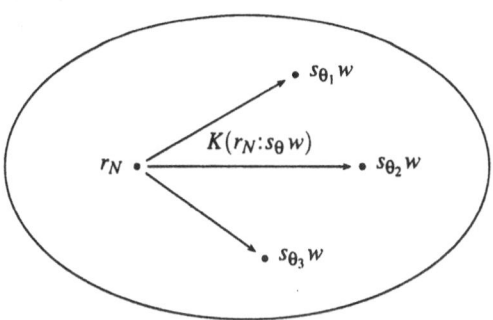

Fig. 2.8. A schematic view of minimum conditional inaccuracy estimation for controlled dynamic systems. The smaller the conditional inaccuracy $\bar{K}(r_N{:}s_\theta)$ of $r_N(y,z)$ relative to $s_\theta(y|z)$ is, the better is the match between data and model. Alternatively, the joint inaccuracy $K(r_N{:}s_\theta w)$ can be used where $w(z)$ is an arbitrary fixed density of Z.

Maximum A Posteriori Estimation. The maximum a posteriori probability (MAP) estimate defined as

$$
\hat{\theta}_{MAP} \triangleq \arg \max_\theta p_N(\theta)
$$

is with respect to the expression (2.34) of the posterior density

$$
p_N(\theta) \propto p_0(\theta) \exp\left(-N\bar{K}(r_N{:}s_\theta)\right)
$$

equivalent to

$$
\hat{\theta}_{MAP} = \arg \min_{\theta \in \mathcal{T}} \left(\bar{K}(r_N{:}s_\theta) - \frac{1}{N} \log p_0(\theta) \right). \tag{2.42}
$$

Compared with the form (2.41) of the ML estimate, the above is modified by the normalized log-prior. When the conjugate prior (2.35) is used

$$p_v(\theta) \propto \exp\left(-v\bar{K}(r_v:s_\theta)\right),$$

the MAP estimate can be put in the following compact form

$$\boxed{\hat{\theta}_{\mathrm{MAP}} = \arg\min_{\theta \in \mathcal{T}} \bar{K}(r_{v+N}:s_\theta).}$$

(2.43)

Bayesian Estimation. In contrast to point estimation, Bayesian estimation aims at calculating the functions

$$\bar{K}(r_N:s_\theta), \quad \theta \in \mathcal{T}$$

or

$$\bar{K}(r_{v+N}:s_\theta), \quad \theta \in \mathcal{T},$$

i.e., conditional inaccuracy of $r_N(y, z)$ or $r_{v+N}(y, z)$, respectively, relative to *all* theoretical densities $s_\theta(y|z)$, $\theta \in \mathcal{T}$. The posterior density is constructed from the above function by means of simple transformations (2.34) and (2.36), respectively.

2.5 Inaccuracy for Discrete Data

When the observed data Y or (Y, Z) are supposed to be discrete random variables taking values in a finite set $\mathcal{Y}$ or $\mathcal{Y} \times \mathcal{Z}$, respectively, the notion of inaccuracy may be given special interpretations that support in turn its use in parameter estimation.

Independent Observations

Empirical Mass Function. Given the observed sample $y_1, \ldots, y_N$, the *empirical mass function* of Y is defined as

$$r_N(y) \overset{\Delta}{=} \frac{1}{N} \sum_{k=1}^{N} \mathbf{1}_{y_k}(y)$$

where $\mathbf{1}_a(y)$ stands for the indicator function

$$\mathbf{1}_a(y) = \begin{cases} 1 & \text{if } y = a, \\ 0 & \text{if } y \neq a. \end{cases}$$

Inaccuracy. Inaccuracy of the empirical mass function $r_N(y)$ relative to a theoretical mass function $s_\theta(y)$ (Kerridge, 1961) is defined through

$$K(r_N:s_\theta) \overset{\Delta}{=} \sum_{y \in \mathcal{Y}} r_N(y) \log \frac{1}{s_\theta(y)}.$$

(2.44)

Relation to Other Information Measures. Inaccuracy $K(r_N:s_\theta)$ is closely linked with *Shannon's entropy* of the empirical mass function $r_N(y)$ (Shannon, 1948)

$$H(r_N) \overset{\Delta}{=} \sum_{y \in \mathcal{Y}} r_N(y) \log \frac{1}{r_N(y)} = K(r_N:r_N)$$

and *Kullback-Leibler (K-L) distance* between the empirical and theoretical mass functions $r_N(y)$ and $s_\theta(y)$, respectively (Kullback and Leibler, 1951)

$$D(r_N \| s_\theta) \overset{\Delta}{=} \sum_{y \in \mathcal{Y}} r_N(y) \log \frac{r_N(y)}{s_\theta(y)}.$$

Indeed, using the above definitions we can write

$$
\begin{aligned}
K(r_N:s_\theta) &= \sum_{y \in \mathcal{Y}} r_N(y) \log \frac{1}{s_\theta(y)} \\
&= \sum_{y \in \mathcal{Y}} r_N(y) \log \frac{1}{r_N(y)} + \sum_{y \in \mathcal{Y}} r_N(y) \log \frac{r_N(y)}{s_\theta(y)} \\
&= H(r_N) + D(r_N \| s_\theta).
\end{aligned}
$$

Hence, we have an interesting identity

$$\boxed{K(r_N:s_\theta) = H(r_N) + D(r_N \| s_\theta).} \tag{2.45}$$

Let us emphasize that the formula does not hold for continuous Y; a formal evaluation gives $H(r_N) = -\infty$ and $D(r_N \| s_\theta) = \infty$ though $K(r_N:s_\theta)$ may still be finite.

When dealing with discrete observations, the identity (2.45) suggests another approach to estimation—via K-L distance. This view is supported by the fact (shown in Chap. 3) that K-L distance is always nonnegative

$$D(r_N \| s_\theta) \geq 0$$

with equality if and only if $s_\theta(y) = r_N(y)$.

Minimum K-L Distance Estimation. Since Shannon's entropy $H(r_N)$ is independent of the unknown parameter θ, it follows from (2.45) that maximizing likelihood and minimizing inaccuracy is equivalent to minimizing K-L distance

$$\hat{\theta}_{ML} = \arg \min_{\theta \in \Theta} D(r_N \| s_\theta)$$

provided the minimum exists.

Example 2.9 *(Probability simplex)* Let Y take on just three different values — 1, 2, 3. Probability mass functions on $\mathcal{Y} = \{1,2,3\}$ can be identified with the probability vectors

$$[\Pr\{Y = 1\}, \Pr\{Y = 2\}, \Pr\{Y = 3\}].$$

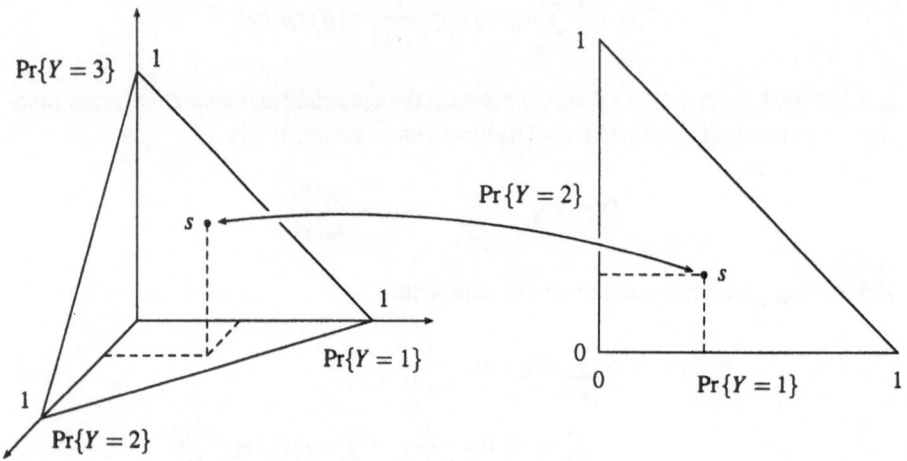

Fig. 2.9. The set of all distributions on $\mathcal{Y} = \{1,2,3\}$ forms a two-dimensional simplex which is a convex envelope of the points $(1,0,0),(0,1,0),(0,0,1)$ in $\mathbb{R}^3$.

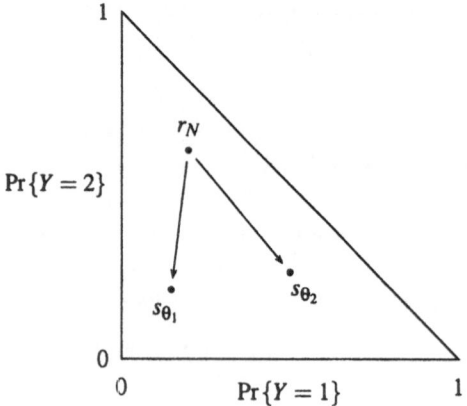

Fig. 2.10. Both the empirical density $r_N(y)$ and the theoretical densities $s_\theta(y)$, $\theta \in \mathcal{T}$ can be regarded as points of the same probability simplex.

Any probability mass function can thus be viewed as a point of the associated probability simplex shown in Fig. 2.9. In this view, maximum likelihood estimation searches for the probability vector s_θ, $\theta \in \mathcal{T}$ that is nearest, in terms of K-L distance, to the probability vector r_N (cf. Fig. 2.10).   □

Example 2.10 (*Binomial distribution*) Consider an experiment with two possible outputs $X = 1$ (success) and $X = 0$ (failure). Let $X = 1$ and $X = 0$ occur with probabilities θ and $1 - \theta$, respectively. We denote through $Y = X_1 + X_2$ the number of occurrences of $X = 1$, i.e., number of successes in a sequence of two trials. The distribution of Y is clearly binomial $B(2, \theta)$ with the mass function

$$s_\theta(y) = \binom{2}{y} \theta^y (1 - \theta)^{2-y}, \quad y \in \{0, 1, 2\}.$$

Suppose that the above two-trial experiment is repeated many times. Fig. 2.12 illustrates on a simulated sample of 100 data that the empirical mass function $r_N(y)$ gets close, for a sufficiently long sample y^N, to the theoretical mass function $s_{\theta_0}(y)$.   □

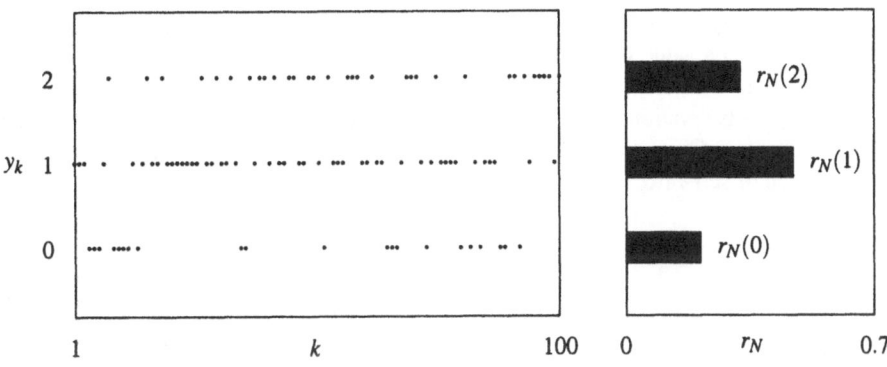

Fig. 2.11. Empirical mass function of 100 data sampled from binomial distribution $B(n, \theta)$ for $n = 2$ and $\theta = 0.6$.

Possible Views of Inaccuracy. There are several appealing interpretations of inaccuracy in the discrete-data case.

Information Theory. The following quantity, often called self-information,

$$\log \frac{1}{s_\theta(y)}$$

can be taken as a simple measure of the amount of information associated with the event $Y = y$, relative to a particular theoretical density $s_\theta(y)$. This interpretation makes a good sense; observing a value y which we assumed beforehand almost certain,

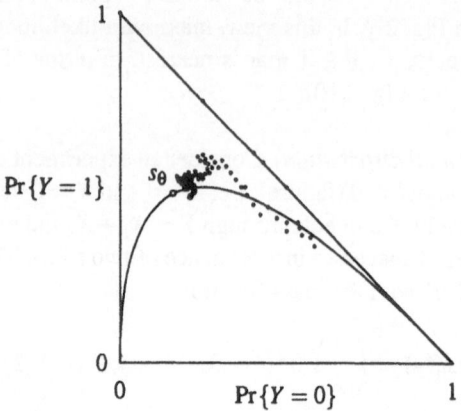

Fig. 2.12. The empirical mass functions r_N (marked by dots) for samples $y^N = (y_1, \ldots, y_N)$ from the binomial distribution $B(n, \theta)$ with $n = 2$ approach the theoretical family (solid curve) near the true point $\theta = 0.6$ (circle) as N becomes large enough.

$s_\theta(y) \approx 1$, brings very little additional information to our knowledge. On the contrary, observing a value y' that we considered almost impossible, $s_\theta(y') \approx 0$, amounts to substantial information that may lead us to correct or reject the current model.

A single observation, due to its random character, is typically not enough to make any conclusions. Given a sample $(y_1, \ldots, y_N)$ we may, however, compute the empirical expectation of self-information

$$E_N\left(\log \frac{1}{s_\theta(Y)}\right) = K(r_N : s_\theta).$$

Taking the above interpretation and the identity (2.45) together, we can view inaccuracy as a measure of total uncertainty of Y. Shannon's entropy $H(r_N)$ measures the intrinsic uncertainty of Y caused by its random behaviour. This component of inaccuracy is "objective"—cannot be influenced by modelling. On the other hand, K-L distance $D(r_N \| s_\theta)$ quantifies the increase of uncertainty due to the use of the theoretical mass function $s_\theta(y)$ to predict Y. To put it another way, K-L distance measures the amount of information that speaks against the model; it quantifies the deviation of $r_N(y)$ from $s_\theta(y)$.

Coding Theory. One of problems solved in information theory is to assign to the events $Y = y$ codewords with length $l(y)$ so to minimize the expected length of code. Using Shannon's code assignment

$$l(y) = \log_2 \frac{1}{s_\theta(y)},$$

we easily conclude that inaccuracy $K(r_N : s_\theta)$ is just the expected length of code

$$E_N(l(y)) = K(r_N : s_\theta).$$

Strictly speaking, the codewords can be of integer length only, therefore the length $l(y)$ must be rounded up to the nearest integer and the above identity holds then with precision up to 1 bit.

Neglecting this point, we can interpret the identity (2.45) as follows. The average length of code for a given sequence $(y_1, \ldots, y_N)$ is a sum of two components: $H(r_N)$ is the minimum length achievable for $s_\theta(y) = r_N(y)$ while $D(r_N \| s_\theta)$ is the increase of the code length due to the use of $s_\theta(y) \neq r_N(y)$ (see, e.g., Cover and Thomas, 1991).

Statistics. From the statistical point of view, inaccuracy (2.44) is owing to (2.23) nothing but negative normalized log-likelihood

$$K(r_N : s_\theta) = -\frac{1}{N} \log l_N(\theta).$$

The minimum inaccuracy achievable within a class of mass functions s_θ is just a transformed value of the maximum likelihood over the class

$$K(r_N : s_{\hat{\theta}_{ML}}) = -\frac{1}{N} \log l_N(\hat{\theta}_{ML}).$$

Probability Theory. With respect to (2.22), inaccuracy, when positive, measures the rate of convergence of the joint probability $q_\theta^N(y^N)$ of sample $(Y_1, \ldots, Y_N)$ to zero

$$K(r_N : s_\theta) = -\frac{1}{N} \log q_\theta^N(y^N).$$

This relationship is usually the starting point for derivation of asymptotic results in large deviation theory (cf. Sect. 4.4).

Controlled Dynamic Systems

Joint Empirical Mass Function. Given the observed samples $y_1, \ldots, y_{N+m}$ and $u_1, \ldots, u_{N+m}$, the joint *empirical mass function* of (Y, Z) is defined as

$$r_N(y, z) \overset{\Delta}{=} \frac{1}{N} \sum_{k=m+1}^{N+m} 1_{y_k, z_k}(y, z)$$

where $1_{a,b}(y, z)$ stands for the indicator function

$$1_{a,b}(y, z) = \begin{cases} 1 & \text{if } y = a \text{ and } z = b, \\ 0 & \text{if } y \neq a \text{ or } z \neq b. \end{cases}$$

Conditional Inaccuracy. Conditional inaccuracy of the joint empirical mass function $r_N(y, z)$ relative to a conditional theoretical mass function $s_\theta(y|z)$ is defined by

$$\tilde{K}(r_N : s_\theta) \overset{\Delta}{=} \sum_{(y,z) \in \mathcal{Y} \times \mathcal{Z}} r_N(y, z) \log \frac{1}{s_\theta(y|z)}.$$

Relation to Other Information Measures. Again, there is a close connection between the conditional inaccuracy $\bar{K}(r_N:s_\theta)$, *conditional Shannon's entropy* of the joint empirical mass function $r_N(y,z)$

$$\bar{H}(r_N) = \sum_{(y,z)\in\mathcal{Y}\times\mathcal{Z}} r_N(y,z) \log \frac{\tilde{r}_N(z)}{r_N(y,z)}$$

and *conditional Kullback-Leibler (K-L) distance* between the joint empirical and conditional theoretical mass functions $r_N(y,z)$ and $s_\theta(y|z)$, respectively

$$\bar{D}(r_N\|s_\theta) = \sum_{(y,z)\in\mathcal{Y}\times\mathcal{Z}} r_N(y,z) \log \frac{r_N(y,z)}{s_\theta(y|z)\,\tilde{r}_N(z)}$$

where

$$\tilde{r}_N(z) = \sum_{y\in\mathcal{Y}} r(y,z)$$

is the marginal empirical mass function. Using the above definitions, we find that

$$\begin{aligned}
\bar{K}(r_N:s_\theta) &= \sum_{(y,z)\in\mathcal{Y}\times\mathcal{Z}} r_N(y,z) \log \frac{1}{s_\theta(y|z)} \\
&= \sum_{(y,z)\in\mathcal{Y}\times\mathcal{Z}} r_N(y,z) \log \frac{\tilde{r}_N(z)}{r_N(y,z)} + \sum_{(y,z)\in\mathcal{Y}\times\mathcal{Z}} r_N(y,z) \log \frac{r_N(y,z)}{s_\theta(y|z)\,\tilde{r}_N(z)} \\
&= \bar{H}(r_N) + \bar{D}(r_N\|s_\theta).
\end{aligned}$$

We obtain thus a "conditional" version of (2.45)

$$\boxed{\bar{K}(r_N:s_\theta) = \bar{H}(r_N) + \bar{D}(r_N\|s_\theta).} \tag{2.46}$$

Note that a similar formula does not hold for continuous (Y,Z); formal evaluation would give $\bar{H}(r) = -\infty$ and $\bar{D}(r\|s) = \infty$.

Similarly as it was for independent observations, the identity (2.46) suggests another approach to estimation—via conditional K-L distance. This view is supported by the fact (shown in Chap. 3 again) that conditional K-L distance is always nonnegative

$$\bar{D}(r_N\|s_\theta) \geq 0$$

with equality if and only if $s_\theta(y|z)\,\tilde{r}_N(z) = r_N(y,z)$.

Minimum Conditional K-L Distance Estimation. Since conditional Shannon's entropy $\bar{H}(r_N)$ is independent of the unknown parameter θ, it follows from (2.46) that maximizing likelihood and minimizing conditional inaccuracy is equivalent to minimizing conditional K-L distance

$$\hat{\theta}_{\mathrm{ML}} = \arg\min_{\theta\in\Theta} \bar{D}(r_N\|s_\theta)$$

provided the minimum exists.

Example 2.11 (*Markov chain*)   Consider a Markov chain with 3 possible states denoted $\{1,2,3\}$. A complete description of the Markov chain is given by the matrix of transition probabilities s_{ij}

		y		
		1	2	3
z	1	s_{11}	s_{12}	s_{13}
	2	s_{21}	s_{22}	s_{23}
	3	s_{31}	s_{32}	s_{33}

where s_{ij} denotes the probability of transition from state i to state j.

The conditional mass function $s(y|z)$ can be regarded as a collection of mass functions of Y parametrized by z

$$s_z(y) = s(y|z), \quad z = 1,2,3.$$

Thus, $s_i \equiv [s_{i1}, s_{i2}, s_{i3}]$. Similarly, the conditional empirical mass function

$$r_N(y|z) = \begin{cases} \dfrac{r_N(y,z)}{r_N(z)} & \text{if } r_N(z) > 0, \\ \text{arbitrary} & \text{if } r_N(z) = 0 \end{cases}$$

can be regarded as a collection of the mass functions

$$r_{z,N}(y) = r_N(y|z), \quad z = 1,2,3.$$

The conditional inaccuracy $\bar{K}(r_N{:}s)$ can be rewritten as the empirical expectation of the inaccuracy of $r_{z,N}(y)$ relative to $s_Z(y)$ (cf. Fig. 2.13)

$$\bar{K}(r_N{:}s) = E_N\big(K(r_{Z,N}{:}s_Z)\big) = \sum_{z=1}^{3} r_N(z)\, K(r_{z,N}{:}s_z).$$

The fact that the conditional empirical mass function $r_N(y|z)$ is not uniquely determined for the values z that have not been observed does not matter here. Since $r_N(z) = 0$ for such z, the ambiguity does not affect the resulting value of $\bar{K}(r_N{:}s)$.

The view may be extended to any discrete-data problem.   □

Possible Views of Inaccuracy. When there are no external inputs to the system considered, the joint probability of sample Y_{m+1}^{N+m} given $Y^m = y^m$ can be expressed by (2.32) as follows

$$q_\theta^N(y_{m+1}^{N+m}|y^m) = \exp\big(-N\bar{K}(r_N{:}s_\theta)\big).$$

This formally concides with the case of independent observations, except that inaccuracy is replaced by its conditional version.

Thus, most interpretations shown above for independent observations hold for dynamic systems as well. For instance, conditional inaccuracy is negative normalized log-likelihood

$$\bar{K}(r_N{:}s_\theta) = -\frac{1}{N} \log l_N(\theta).$$

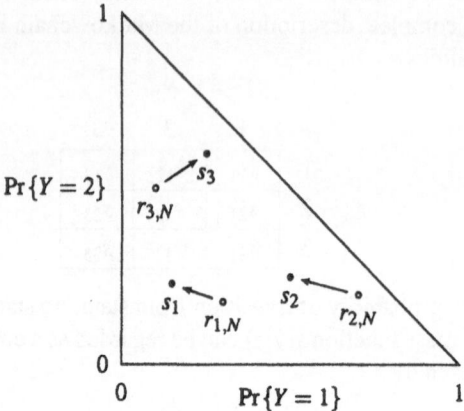

Fig. 2.13. Conditional inaccuracy $\bar{K}(r_N:s_\theta)$ can be regarded as a sample average of the inaccuracies $K(r_{z,N}:s_z)$ of $r_{z,N}(y)$ relative to $s_z(y)$ for $z = 1, 2, 3$.

The minimum conditional inaccuracy achievable within a class of mass functions $s_\theta(y|z)$ is just a transformed value of the maximum likelihood over the class

$$\bar{K}(r_N:s_{\theta_{ML}}) = -\frac{1}{N} \log l_N(\hat{\theta}_{ML}).$$

Conditional inaccuracy, when positive, measures the rate of convergence of the joint probability $q_\theta^N(y_{m+1}^{N+m}|y^m)$ to zero

$$\bar{K}(r_N:s_\theta) = -\frac{1}{N} \log q_\theta^N(y_{m+1}^{N+m}|y^m).$$

When external inputs are present, the situation is more difficult. The joint probability of sample (2.32)

$$q_\theta^N(y_{m+1}^{N+m}, u_{m+1}^{N+m}|y^m, u^m) = \Gamma(y^{N+m}, u^{N+m}) \exp(-N\bar{K}(r_N:s_\theta))$$

depends obviously on the distribution of inputs. In applications, however, the distribution of external inputs is rarely modelled and identified. Luckily, under natural conditions of control (2.15), the factor

$$\Gamma(y^{N+m}, u^{N+m}) = \prod_{k=m+1}^{N+m} \gamma_k(u_k|y^{k-1}, u^{k-1})$$

is independent of θ and thus may be treated as a constant for a given sample. This greatly simplifies the modelling requirements but we pay for this by the lack of symmetry with the completely described problems.

All we can say then is that conditional inaccuracy is equal, *up to an additive constant C*, to negative normalized log-likelihood

$$\bar{K}(r_N:s_\theta) = C - \frac{1}{N} \log l_N(\theta).$$

2.6 Examples: Probability Matching

The following examples aim at giving a better feel for the idea of probability matching and the particular role of inaccuracy and conditional inaccuracy in it.

Independent Observations

Example 2.12 (*Bernoulli distribution*) Consider a simple model of coin tossing where $\mathcal{Y} = \{\text{Head}, \text{Tail}\}$ and $s_\theta(y)$ is θ if $y = \text{Head}$ and $1 - \theta$ if $y = \text{Tail}$. Let $\hat{\theta}_N$ be the relative frequency of heads observed in the sequence of trials y^N. The inaccuracy of the corresponding probability vectors is then

$$K(r_N : s_\theta) = K\big([\hat{\theta}_N, 1 - \hat{\theta}_N] \| [\theta, 1 - \theta]\big)$$
$$= H\big([\hat{\theta}_N, 1 - \hat{\theta}_N]\big) + D\big([\hat{\theta}_N, 1 - \hat{\theta}_N] \| [\theta, 1 - \theta]\big).$$

It is not difficult to see that

$$K(r_N : s_\theta) - K(r_N : s_{\hat{\theta}_N}) = D\big([\hat{\theta}_N, 1 - \hat{\theta}_N] \| [\theta, 1 - \theta]\big) \geq 0.$$

Thus, the minimum of inaccuracy on $\mathcal{T}$ is attained at $\theta = \hat{\theta}_N$.   □

Example 2.13 (*Normal distribution*) Let Y be normally distributed with an unknown mean θ and known, constant variance σ^2, $Y \sim N(\theta, \sigma^2)$. Straightforward calculations yield inaccuracy

$$K(s_{\theta'} : s_\theta) = \frac{1}{2} \log 2\pi\sigma^2 + \frac{1}{2} + \frac{1}{2\sigma^2}(\theta' - \theta)^2.$$

By (2.45), inaccuracy can be decomposed as

$$K(s_{\theta'} : s_\theta) = H(s_{\theta'}) + D(s_{\theta'} \| s_\theta)$$

using Shannon's entropy

$$H(s_{\theta'}) = \int s_{\theta'}(y) \log \frac{1}{s_{\theta'}(y)}\, dy = \frac{1}{2} \log 2\pi\sigma^2 + \frac{1}{2}$$

and K-L distance

$$D(s_{\theta'} \| s_\theta) = \int s_{\theta'}(y) \log \frac{s_{\theta'}(y)}{s_\theta(y)}\, dy = \frac{1}{2\sigma^2}(\theta' - \theta)^2.$$

Note that the minimum of both the inaccuracy and K-L distance is attained at $\theta' = \theta$.   □

Example 2.14 (*Geyser eruptions*) A sample of 107 values $(y_1, \ldots, y_{107})$ shown in Fig. 2.14 records the lengths of eruptions of the Old Faithful geyser in Yellowstone National Park[1].

---

[1] The data can be found in Weisberg (1980) or Silverman (1986).

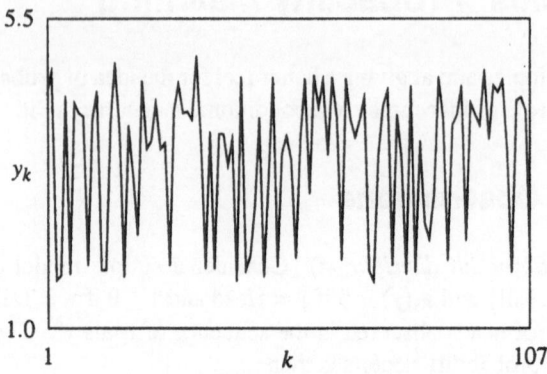

Fig. 2.14. Geyser eruptions: eruption length (in minutes) of Old Faithful geyser.

The view of the problem via the empirical density $r_N(y)$, indicated in Fig. 2.15 and 2.16 by histograms, gives immediately more insight into the actual distribution of data. In particular, it reveals its bimodal character; the sample is a mixture of shorter and longer eruptions.

One possible model of data is to define the theoretical density $s_\theta(y)$ as a weighted mixture of normal densities

$$s_\theta(y) = (1 - \theta_5)\, s_{\theta_1, \theta_3}(y) + \theta_5\, s_{\theta_2, \theta_4}(y)$$

where s_{θ_1, θ_3} and s_{θ_2, θ_4} are normal densities with means θ_1 and θ_2 and variances θ_3 and θ_4, respectively. The minimum inaccuracy (maximum likelihood) estimate of θ gives a very good agreement between model and data (see Fig. 2.15).

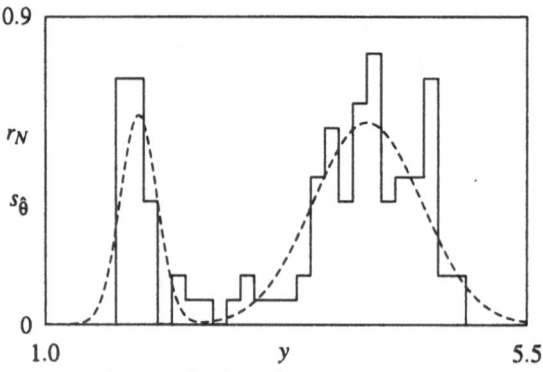

Fig. 2.15. Geyser eruptions: histogram of data (solid line) versus the minimum inaccuracy weighted sum of normal densities (dashed line).

Another possibility is to fit the empirical density $r_N(y)$ with an exponential of a fourth-order polynomial in y (cf. Barron and Sheu, 1991)

$$s_\theta(y) = \frac{\exp\left(\sum_{i=1}^4 \theta_i y^i\right)}{\int_{\mathbb{R}} \exp\left(\sum_{i=1}^4 \theta_i y^i\right) dy}.$$

The fit attained with this model (see Fig. 2.16) looks worse than in the preceding case but the parameters of the exponential are much easier to estimate recursively as we shall see in the next chapter.   □

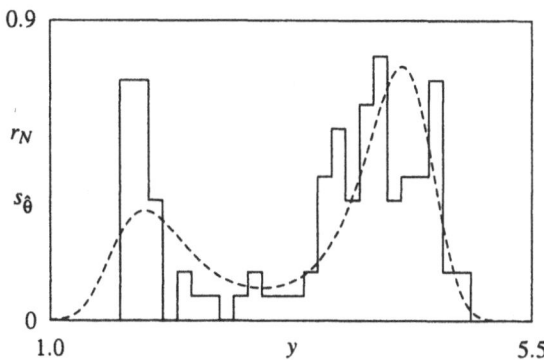

Fig. 2.16. Geyser eruptions: histogram of data (solid line) versus the minimum inaccuracy exponential of a 4th-order polynomial (dashed line).

Example 2.15 (*Horsekick deaths*) Table 2.1 shows the numbers of deaths caused by horsekicks in 14 corps of the Prussian Army in 1875–1894[2].

To model rarely occurring events like this, Poisson distribution is a common choice. The Poisson mass function is defined as

$$s_\theta(y) = \frac{\exp(-\theta)\,\theta^y}{y!}, \quad y = 0, 1, 2, \ldots$$

where $y!$ denotes the factorial of y defined by $y! = y(y-1)\cdots 1$ for $y > 1$ with $0! = 1$.

The fact that we have 14 different samples available allows us to analyse the sensitivity of parameter estimation to the change of sample. Figure 2.17 shows the inaccuracies

$$K(r_{N,j}:s_\theta) = \theta - \left(\frac{1}{N}\sum_{k=1}^N y_{k,j}\right)\log\theta + \frac{1}{N}\log\left(\prod_{k=1}^N y_{k,j}!\right)$$

as functions of θ for the empirical mass functions $r_{N,j}(y)$ corresponding to particular samples (columns in Table 2.1) $y_j = (y_{1,j},\ldots,y_{20,j})$ for $j = 1,\ldots,14$. Note that the samples were relatively short given $N = 20$.

---

[2]The original data file is available as *http://lib.stat.cmu.edu/datasets/Andrews/T04.1* from StatLib.

Table 2.1. Horsekick deaths in 14 corps of the Prussian Army in 1875–1894.

year	corps no.													
	1	2	3	4	5	6	7	8	9	10	11	12	13	14
1875	0	0	0	0	0	0	0	1	1	0	0	0	1	0
1876	2	0	0	0	1	0	0	0	0	0	0	0	1	1
1877	2	0	0	0	0	0	1	1	0	0	1	0	2	0
1878	1	2	2	1	1	0	0	0	0	0	1	0	1	0
1879	0	0	0	1	1	2	2	0	1	0	0	2	1	0
1880	0	3	2	1	1	1	0	0	0	2	1	4	3	0
1881	1	0	0	2	1	0	0	1	0	1	0	0	0	0
1882	1	2	0	0	0	0	1	0	1	1	2	1	4	1
1883	0	0	1	2	0	1	2	1	0	1	0	3	0	0
1884	3	0	1	0	0	0	0	1	0	0	2	0	1	1
1885	0	0	0	0	0	0	1	0	0	2	0	1	0	1
1886	2	1	0	0	1	1	1	0	0	1	0	1	3	0
1887	1	1	2	1	0	0	3	2	1	1	0	1	2	0
1888	0	1	1	0	0	1	1	0	0	0	0	1	1	0
1889	0	0	1	1	0	1	1	0	0	1	2	2	0	2
1890	1	2	0	2	0	1	1	2	0	2	1	1	2	2
1891	0	0	0	1	1	1	0	1	1	0	3	3	1	0
1892	1	3	2	0	1	1	3	0	1	1	0	1	1	0
1893	0	1	0	0	0	1	0	2	0	0	1	3	0	0
1894	1	0	0	0	0	0	0	0	1	0	1	1	0	0

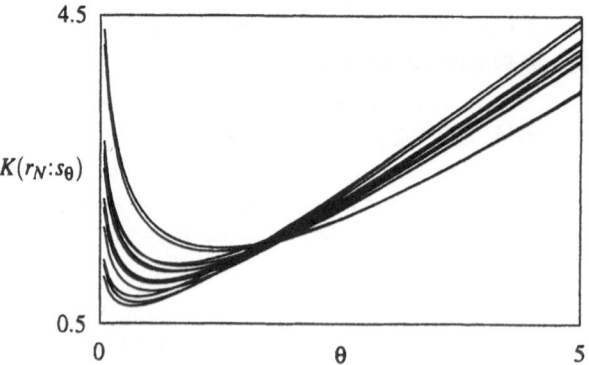

Fig. 2.17. Horsekick deaths: inaccuracies for particular corps as functions of the parameter θ of Poisson distribution.

To illustrate a good agreement between the empirical and theoretical mass functions achieved for the minimum inaccuracy estimate (2.38), Fig. 2.18 compares the values of $r_{N,j}(y)$ and $s_{\hat{\theta}}(y)$ for $y = 0, 1, 2, 3, 4$ and corps no. $j = 1$.   □

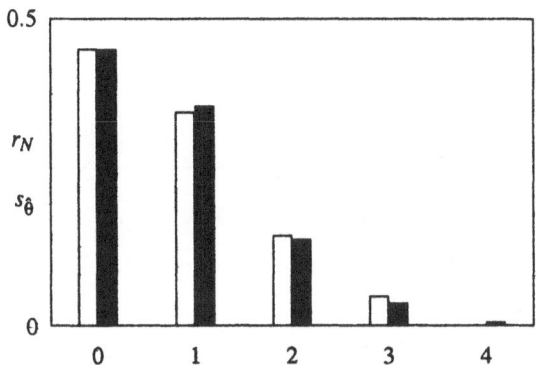

Fig. 2.18. Horsekick deaths: agreement between the empirical mass function (white bar graph) and Poisson mass function (shaded bar graph) for the minimum inaccuracy estimate $\hat{\theta} = 0.8$.

Example 2.16 (*Balloon measurements*)   The data consist of 2,001 measurements of radiation taken from a balloon about 30–40 kilometers above the earth's surface. The measuring device was suspended from the balloon by ropes. Due to rotation of the balloon, the direct radiation from the sun was occasionally cut off by the ropes, giving rise to outliers. The original data exhibit a trend as the amount of radiation increases with altitude. Here only residuals after removing the trend are considered[3], see Fig. 2.19.

The empirical density shown by histogram in Fig. 2.20 suggests to be approximated with a mixture of a normal distribution around zero and a skew heavy-tailed distribution over negative values. One possible definition of the theoretical density is to take the mixture

$$s_\theta(y) = \varepsilon s_{\alpha,\beta}(y) + (1 - \varepsilon) s_{\gamma,\delta,\nu}(y)$$

where the normal density

$$s_{\alpha,\beta}(y) = \frac{1}{\sqrt{2\pi\beta}} \exp\left(-\frac{1}{2\beta}(y - \alpha)^2\right)$$

may be seen as explanation of the atmospheric variations while the log-Student density over the negative values of y

---

[3]The original data file is available as *http://lib.stat.cmu.edu/datasets/balloon* from StatLib. The balloon measurements were used by Davies and Gather (1993) to exemplify the appearance of outliers in real-life data.

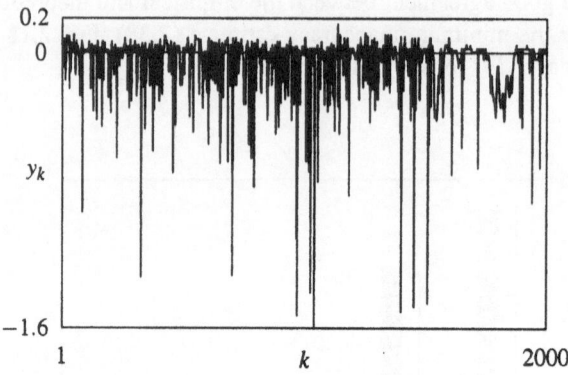

Fig. 2.19. Balloon data: measurement residuals.

$$s_{\gamma,\delta,\nu}(y) = \begin{cases} \dfrac{1}{\sqrt{\pi\nu\delta}} \dfrac{\Gamma(\frac{\nu+1}{2})}{\Gamma(\frac{\nu}{2})} \dfrac{1}{y}\left(1 + \dfrac{(\log(-y)-\log\gamma)^2}{\nu\delta}\right)^{-\frac{\nu+1}{2}} & \text{if } y < 0, \\ \\ 0 & \text{if } y \geq 0 \end{cases}$$

models the effect of rope shading.

Altogether, the unknown parameters are

$$\theta = (\alpha, \beta, \gamma, \delta, \nu, \varepsilon)$$

The minimum inaccuracy (maximum likelihood) estimate of the parameter vector θ gives a very good agreement between $r_N(y)$ and $s_{\hat{\theta}}(y)$ both in the central part around zero (see Fig. 2.20) and in the tails (see Fig. 2.21). To be able to envisage the empirical density in semilogarithmic coordinates, we substituted for $r_N(y)$ its smoothened version given by a weighted sum of normal densities located at particular observed values y_i, with fixed variance $\sigma^2 = 0.0005$.

The minimum inaccuracy estimate of θ was found found by numerical optimization. The values found were

$$\hat{\alpha} = 0.0097, \ \hat{\beta} = 0.0009, \ \hat{\gamma} = -0.1842,$$
$$\hat{\delta} = 0.4070, \ \hat{\nu} = 8.9271, \ \hat{\varepsilon} = 0.7725.$$

The sensitivity of the inaccuracy $K(r_N{:}s_\theta)$ to the change of particular parameters from the optimum is shown in Fig. 2.22. We use here the notation

$$K(r_N{:}s_\alpha) = K(r_N{:}s_{\alpha,\hat{\beta},\hat{\gamma},\hat{\delta},\hat{\nu},\hat{\varepsilon}})$$

etc. The plots are deliberately chosen so that their vertical scales are common for all parameters. This helps to assess quickly how precisely particular parameters need to be specified. Thus, for instance, the variance β of the normal component is really critical here whereas the number of degrees of freedom ν of the log-Student distribution does not matter too much. □

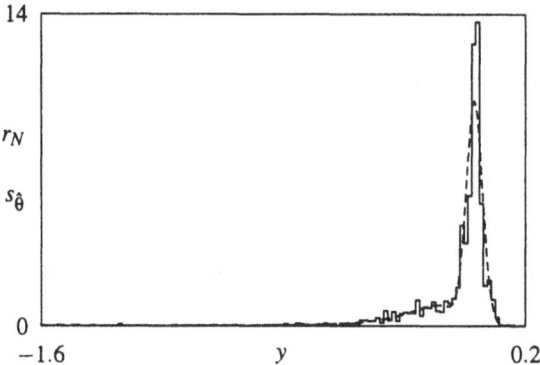

Fig. 2.20. Balloon data: the minimum inaccuracy theoretical density (dashed line) compared with the empirical density envisaged with a histogram (solid line).

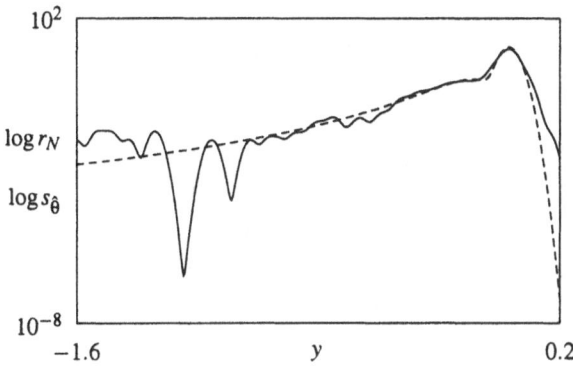

Fig. 2.21. Balloon data: the minimum inaccuracy theoretical density (dashed line) compared with the empirical density smoothened using Gaussian kernels with $\sigma^2 = 0.0005$ (solid line). The vertical scale is logarithmic.

Controlled Dynamic Systems

Generally speaking, it is more difficult to get insight into samples (y^{N+m}, z^{N+m}) produced by controlled dynamic systems. The problem is the dimension of the underlying space of (y, z) which for single-input single-output systems is typically 4–10 but it may be much higher if more variables are involved or the variables are multivariable. It is fair to say that the intuitiveness of the empirical distribution is quickly lost in higher dimensions (as with any other method). We point out some ways how at least partial insight can be gained in higher dimension.

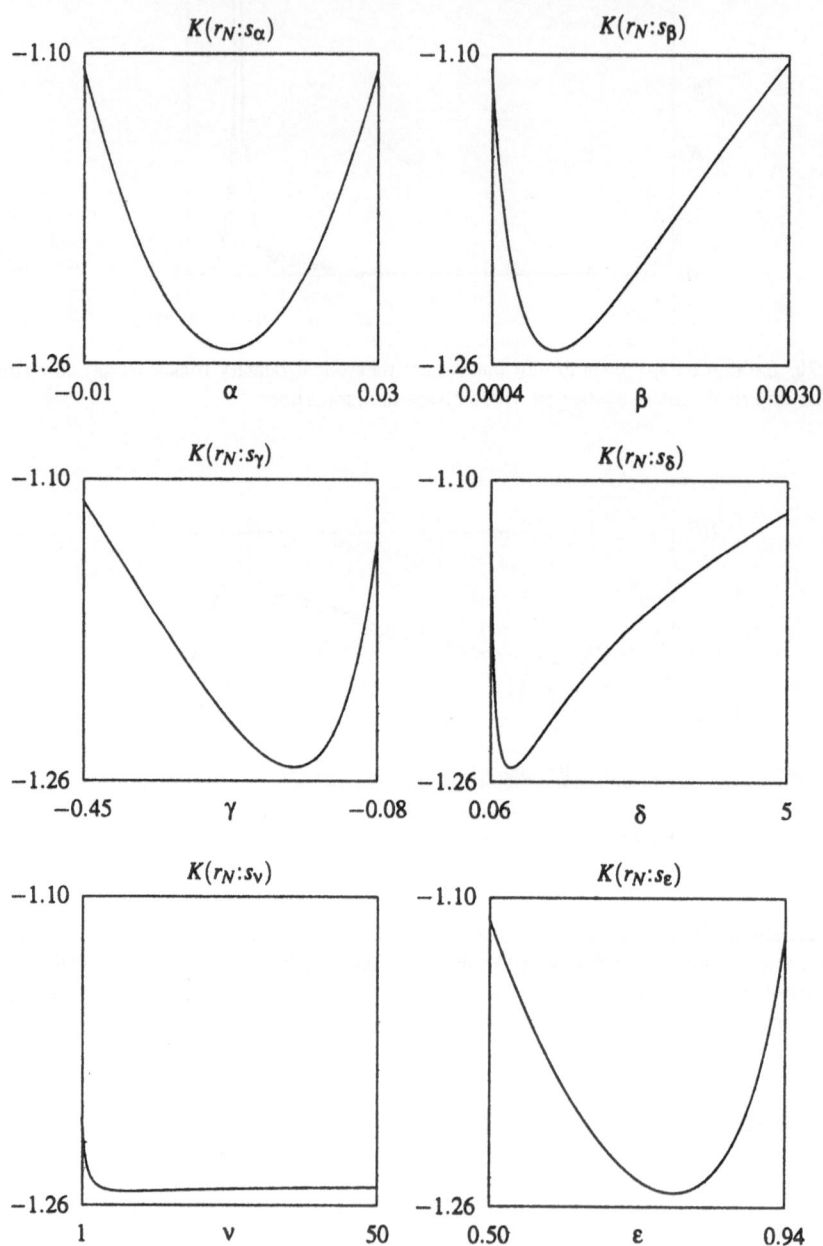

Fig. 2.22. Balloon data: sensitivity of inaccuracy to particular parameters of the theoretical density.

Example 2.17 (*Markov chain*) Consider a simple model of "weather forecast" where Y_k denotes weather on k-th day, $Z_k = Y_{k-1}$, $\mathcal{Y} = \mathcal{Z} = \{$Sunny, Rainy$\}$, $s_\theta(y|z) = \theta$ if $y = z$ and $1 - \theta$ if $y \neq z$. The problem is thus described by the matrix of transition probabilities

	sunny tomorrow	rainy tomorrow
sunny today	θ	$1 - \theta$
rainy today	$1 - \theta$	θ

where θ stands for the probability of steady weather.

Let σ_N be the relative frequency of sunny days followed by another sunny day, ρ_N be the relative frequency of rainy days followed by another rainy day, Σ_N be the relative frequency of sunny days followed by any day. A few seconds of thinking gives

$$\begin{aligned}
\bar{K}(r_N{:}s_\theta) &= \Sigma_N K\big([\sigma_N, 1 - \sigma_N] \| [\theta, 1 - \theta]\big) \\
&\quad + (1 - \Sigma_N) K\big([\rho_N, 1 - \rho_N] \| [\theta, 1 - \theta]\big) \\
&= K\big([\hat{\theta}_N, 1 - \hat{\theta}_N]{:}[\theta, 1 - \theta]\big) \\
&= H\big([\hat{\theta}_N, 1 - \hat{\theta}_N]\big) + D\big([\hat{\theta}_N, 1 - \hat{\theta}_N] \| [\theta, 1 - \theta]\big)
\end{aligned}$$

where

$$\hat{\theta}_N = \Sigma_N \sigma_N + (1 - \Sigma_N)\rho_N.$$

The obvious inequality

$$\bar{K}(r_N{:}s_\theta) - \bar{K}(r_N{:}s_{\hat{\theta}_N}) = D\big([\hat{\theta}_N, 1 - \hat{\theta}_N] \| [\theta, 1 - \theta]\big) \geq 0.$$

implies that $\theta = \hat{\theta}_N$ minimizes inaccuracy on $\mathcal{T}$. □

Example 2.18 (*Sunspot numbers*) Figure 2.23 shows monthly sunspot numbers as observed in Zurich in 1749–1983[4]. The time series exhibits a nonlinear and periodic behaviour which makes it a popular benchmark problem.

Assuming the simplest possible dependence structure with $z_k = y_{k-1}$, the corresponding empirical density $r_N(y, z)$ can be well envisaged using a scatter plot in Fig. 2.24.

Another possibility how to visualize data is to construct estimates of the conditional empirical densities $r_N(y|z)$ for particular values of z. The view of the joint density $r_N(y, z)$ through the conditional cross-sections $r_N(y|z)$ is attractive because of the massive drop in dimension of the underlying space. The difficulty is that any such estimate requires the discrete empirical distribution be smoothened first in some way. The simplest solution is perhaps to take together all observations y_k with z_k belonging to a certain interval I_i; the size of the interval controls the degree of smoothing.

In Fig. 2.25, four such intervals are indicated by the shaded areas. All points that lie within these areas are taken as belonging together and they are used for estimate of the distribution of Y. This is done in Fig. 2.26 through histograms of those points. □

---

[4] The original data file is available as *http://lib.stat.cmu.edu/datasets/Andrews/T11.1* from StatLib.

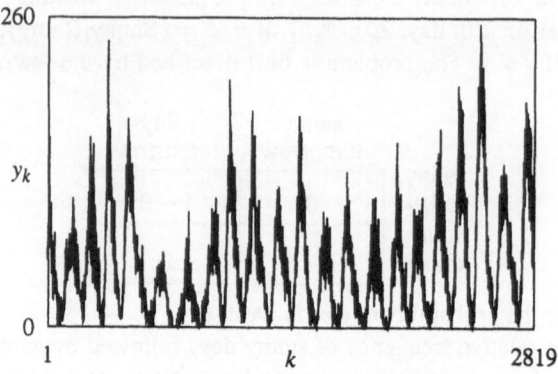

Fig. 2.23. Sunspot numbers: monthly observations in 1749–1983.

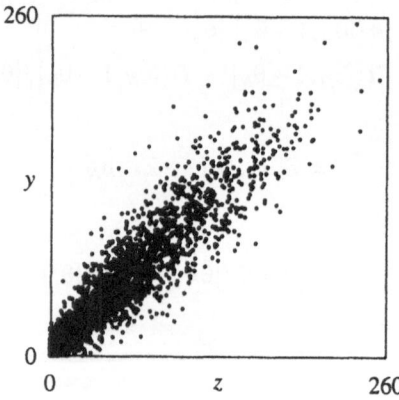

Fig. 2.24. Sunspot numbers: the scatter plot of (y_k, z_k) with $z_k = y_{k-1}$ indicates the joint empirical density of (Y, Z).

Example 2.19 (*Hydraulic actuator*) The data sample shown in Fig. 2.27 concerns a hydraulic actuator that controls the position of a robot arm[5]. The oil pressure in the actuator is affected by the size of the valve opening through which the oil flows into the actuator. The position of the robot arm is then a function of the oil pressure. The oil pressure exhibits an oscillatory behaviour due to mechanical resonances in the robot arm (cf. the lower plot in Fig. 2.27). There is no visible randomness in the behaviour of the system, the difficulty in identifying the system is purely in its higher-order and nonlinear dynamics. The system was identified by Sjöberg (1995) using various approaches.

---

[5]The data were kindly supplied by Jonas Sjöberg, Div. of Autom. Control, Dept. of Electrical Engineering, Linköping University.

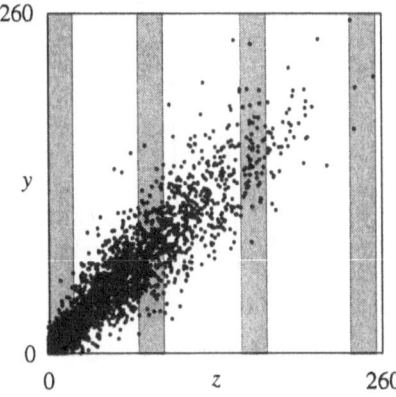

Fig. 2.25. Sunspot numbers: restricting to pairs (y, z) with z belonging to selected intervals I_i (shaded areas) gives an idea about the conditional empirical densities $r_N(y|z)$ for $z \in I_i$ (cf. Fig. 2.26).

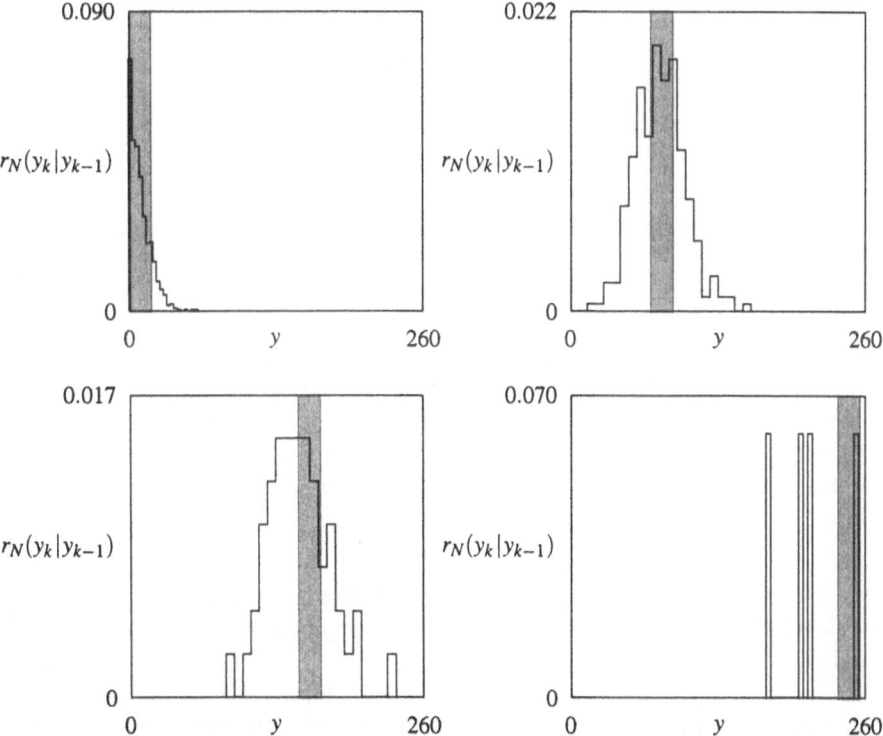

Fig. 2.26. Sunspot numbers: illustration of the conditional empirical densities $r_N(y|z)$ using histograms of those y_k for which z_k belongs to selected intervals I_i (shaded areas). Cf. Fig. 2.25.

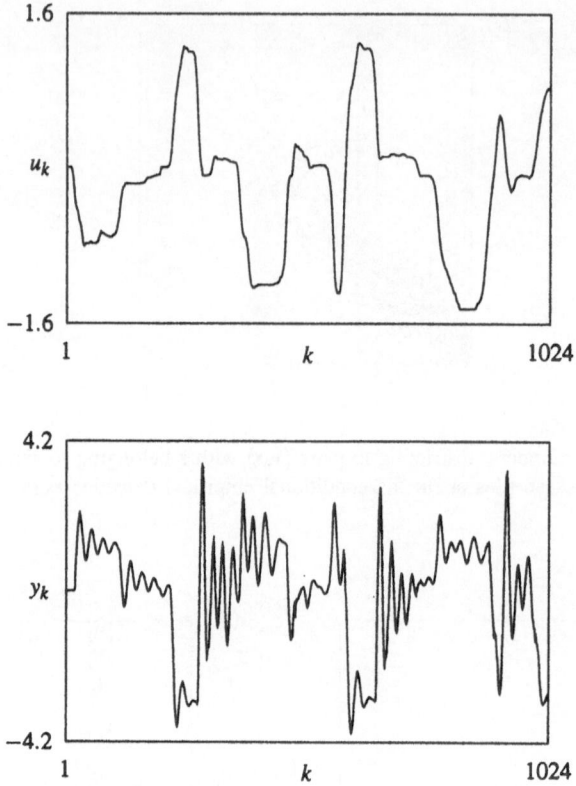

Fig. 2.27. Hydraulic actuator: valve position (input) and oil pressure (output).

The scatter plot of (y_k, u_k) gives a good idea about the empirical density $r_N(y, u)$. It clearly demonstrates that the system is dynamic, i.e., the behaviour of the output y_k is affected not only by the current input u_k but by past data as well.

The same observation can be made using the estimates of the conditional empirical densities $r_N(y|z)$. We used the following procedure for the estimate. First we found all vectors z_i among $z_1, \ldots, z_N$ that were close to the vector z_{50} in terms of the Euclidean distance

$$\|z_i - z_{50}\| < 0.2.$$

Then we used the corresponding y_i points to construct a histogram approximating the conditional empirical density $r_N(y|z = z_{50})$. The same procedure was applied for z_{150}, z_{250} and z_{350}. The results are shown in Fig. 2.29.

For comparison, in the same figure we show the theoretical conditional density $s_{\hat{\theta}}(y|u)$ obtained by the least-squares fit of linear normal regression to data. Clearly, the fit attained is not very good. We could try to make the fit better using more sophisticated models but what cannot be improved is the large variability of Y itself. This is a consequence of the inappropriate structural choice $z_k = u_k$.

To do better, we used a third-order model with

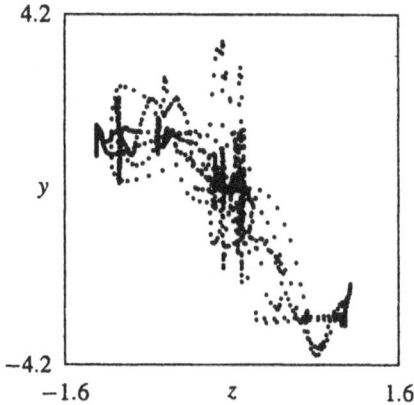

Fig. 2.28. Hydraulic actuator: the scatter plot of (y_k, z_k) with $z_k = u_k$ indicates the joint empirical density of (Y, Z).

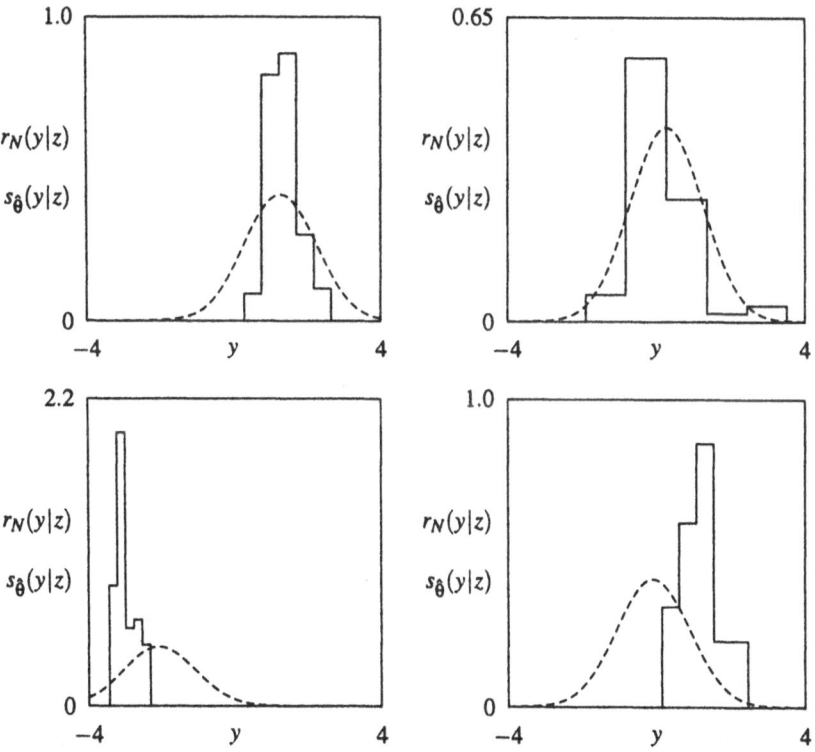

Fig. 2.29. Hydraulic actuator: comparison of the conditional empirical densities envisaged by histograms (solid lines) and theoretical densities (dashed lines) for zero-order static model $y_k = \theta z_k + e_k$ with $z_k = u_k$.

$$z_k = \left[u_k, y_{k-1}, u_{k-1}, y_{k-2}, u_{k-2}\right]^T$$

and applied the same procedure as above. The comparison of histograms approximating the conditional empirical density with the theoretical density obtained by the best linear fit is done in Fig. 2.30.

The fit is much better and the variability of Y dropped significantly. The uncertainty of our estimate increased, however, as well. While the histograms in Fig. 2.29 were constructed from $123, 328, 72$ and 424 data points for $k = 50, 150, 250, 350$, respectively, the histograms in Fig. 2.30 are based on $20, 31, 12, 5$ data points only. Clearly, as the number of all data points (y_k, z_k) is fixed, when N points are put in a higher-dimensional space, they typically become more distant from each other. This warns us that some trade-off between match of data and complexity of model is necessary. □

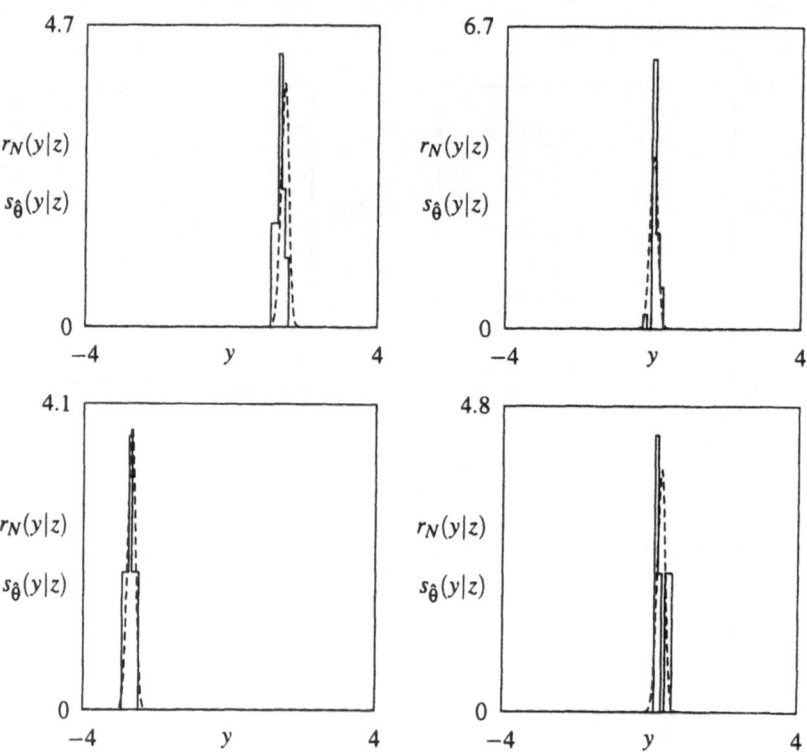

Fig. 2.30. Hydraulic actuator: comparison of the conditional empirical densities envisaged by histograms (solid lines) and theoretical densities (dashed lines) for third-order linear normal regression.

2.7 Historical Notes

Inaccuracy. The concept of inaccuracy was introduced explicitly and analysed in detail in Kerridge (1961). The axiomatic characterization of inaccuracy was given in Kannappan (1972b; 1972a).

Inaccuracy appears quite naturally in information theory as the expected length of Shannon code designed for a theoretical distribution different from the true one (Shannon, 1948). A readable introduction into this point can be found in Cover and Thomas (1991, Sect. 5.4) and Rissannen (1989, Sect. 2.2).

Minimum Distance Estimation. The idea to view estimation as calculation of a certain distance between the empirical and theoretical distributions of data dates back to Wolfowitz (1957). He analysed consistency of minimum distance estimates for a large class of distances.

In robust statistics, *D*-estimators appeared as an alternative way of approaching design of robust estimators, see Beran (1977), Parr and Schucany (1980), Boos (1981), Vajda (1982). The connection between the choice of a particular distance and the estimator robustness was studied intensively. Vajda (1984c; 1984b; 1984a) developed a general theory of *D*-estimators (see also Vajda, 1989).

In system identification, information distances were applied most notably in structure determination. Akaike (1973; 1974) used the concepts of inaccuracy and Kullback-Leibler distance in derivation of the well-known AIC information criterion. The view of estimation as a minimum distance problem was also the starting point for design of approximate estimation and filtering in Hanzon (1987), Stoorvogel and van Schuppen (1995), Kulhavý (1995b).

Connection with Maximum Likelihood. The fact that the minimum inaccuracy (or Kullback-Leibler distance) estimate coincides with the maximum likelihood estimate has been pointed out by many authors, among others Kullback (1959), Kriz and Talacko (1968), Hartigan (1967), Čencov (1972), Akaike (1973).

3. Optimal Estimation with Compressed Data

The view of estimation as 'probability matching' centres around the notion of the empirical density. To determine the empirical density, one basically needs to have access to the complete sample. This is a serious limitation, of course, especially in recursive estimation where the amount of data grows with time. In cases like that, a condensed description of data is required.

When the parametric family of densities is of special—exponential type, estimation can be organized in a recursive form, using only a finite-dimensional statistic of past data. This fact, well known in statistical theory, is given here an appealing geometric interpretation. The geometric tools developed in this connection will serve us in Chap. 4 for design of approximate estimation in families for which the sufficient statistic has a too large or even infinite dimension.

3.1 Data Compression in Euclidean Space

This section continues the Euclidean analogy introduced in Sect. 2.1. Many features of the Pythagorean geometry in Euclidean space will be shown in the following sections to have a natural counterpart in spaces of density functions.

We show that if all model points $\hat{y}^N(\theta)$, $\theta \in \mathcal{T}$ happen to lie within an n-dimensional hyperplane $\mathcal{H}$ in $\mathbb{R}^N$, then the minimum Euclidean distance projection of y^N onto $\mathcal{H}$ is sufficient to reconstruct the "error function" $\|y^N - \hat{y}^N(\theta)\|^2$ with precision up to an additive constant.

Uncorrelated Data

Orthogonal Projection. The least-squares matching of the sequence of observed data $y^N = (y_1, \ldots, y_N)$ with a constant sequence $\theta 1^N$ according to the model

$$y_k = \theta + e_k, \quad k = 1, \ldots, N$$

minimizes the Euclidean distance squared between both the vectors

$$\min_{\theta} \|y^N - \theta 1^N\|^2.$$

A necessary condition for the optimum $\hat{\theta}_N$ follows by differentiating with respect to θ

$$\frac{d}{d\theta} \left\| y^N - \hat{\theta}_N 1^N \right\|^2 = -2 \left(y^N - \hat{\theta}_N 1^N \right) \left(1^N \right)^T = 0.$$

This implies the following condition

$$\left(y^N - \hat{\theta}_N 1^N \right) \left(1^N \right)^T = 0 \tag{3.1}$$

which can be read so that the vector $y^N - \hat{\theta}_N 1^N$ is *orthogonal* (with respect to the inner product $u^T v = \sum_{i=1}^n u_i v_i$) to the basis vector 1^N (see Fig. 3.1). In other words, $\hat{\theta}_N 1^N$ is an orthogonal projection of y^N onto the straight line $\{\theta 1^N : \theta \in \mathbb{R}\}$. The solution to (3.1) is

$$\hat{\theta}_N = \frac{\left(y^N \right) \left(1^N \right)^T}{\left(1^N \right) \left(1^N \right)^T} = \frac{\sum_{k=1}^N y_k}{N} = \frac{1}{N} \sum_{k=1}^N y_k. \tag{3.2}$$

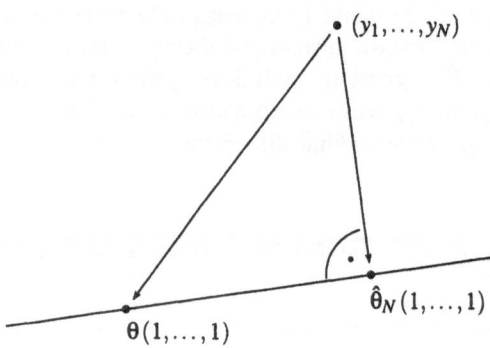

Fig. 3.1. The point $\hat{\theta}_N (1, \ldots, 1)$ minimizing the Euclidean distance between the point $(y_1, \ldots, y_N)$ and the straight line $\{\theta (1, \ldots, 1) : \theta \in \mathbb{R}\}$ is the orthogonal projection of y^N onto the straight line.

Pythagorean Relationship. To prove that $\hat{\theta}_N$ is even the global minimum, we first rewrite

$$\begin{aligned}
\left\| y^N - \theta 1^N \right\|^2 &= \left\| y^N - \hat{\theta}_N 1^N + \hat{\theta}_N 1^N - \theta 1^N \right\|^2 \\
&= \left\| y^N - \hat{\theta}_N 1^N \right\|^2 + \left\| \hat{\theta}_N 1^N - \theta 1^N \right\|^2 \\
&\quad + 2 \left(y^N - \hat{\theta}_N 1^N \right)^T \left(\hat{\theta}_N 1^N - \theta 1^N \right).
\end{aligned}$$

Taking into account (3.1), we see that the orthogonal projection $\hat{\theta}_N 1^N$ satisfies the *Pythagorean relationship*

$$\left\| y^N - \theta \, 1^N \right\|^2 = \left\| y^N - \hat{\theta}_N \, 1^N \right\|^2 + \left\| \hat{\theta}_N \, 1^N - \theta \, 1^N \right\|^2. \qquad (3.3)$$

The identity says that the Euclidean distance squared between y^N and any point $\theta \, 1^N$ is equal to a sum of the Euclidean distance squared between y^N and the orthogonal projection $\hat{\theta}_N \, 1^N$ and the Euclidean distance squared between $\hat{\theta}_N \, 1^N$ and $\theta \, 1^N$. The *global minimum* property easily follows from the Pythagorean relationship: since for every θ

$$\left\| \hat{\theta}_N \, 1^N - \theta \, 1^N \right\|^2 \geq 0,$$

we have

$$\left\| y^N - \theta \, 1^N \right\|^2 \geq \left\| y^N - \hat{\theta}_N \, 1^N \right\|^2$$

with equality if and only if $\theta = \hat{\theta}_N$.

Uniqueness of Projection. The *uniqueness* of the orthogonal projection is also easy to prove. Suppose that two different points $\hat{\theta} \, 1^N$ and $\hat{\theta}' \, 1^N$, $\hat{\theta} \neq \hat{\theta}'$ satisfy (3.1). By Pythagorean relationship (3.3) it must hold

$$\left\| y^N - \hat{\theta} \, 1^N \right\|^2 \geq \left\| y^N - \hat{\theta}' \, 1^N \right\|^2$$

and, at the same time,

$$\left\| y^N - \hat{\theta}' \, 1^N \right\|^2 \geq \left\| y^N - \hat{\theta} \, 1^N \right\|^2.$$

Consequently, it must be

$$\left\| y^N - \hat{\theta}' \, 1^N \right\|^2 = \left\| y^N - \hat{\theta} \, 1^N \right\|^2$$

but the equality appears if and only if $\hat{\theta}' = \hat{\theta}$ which is contradiction with the assumption. Therefore, the orthogonal projection is unique.

Data Compression. Note that the Euclidean distance squared $\left\| y^N - \hat{\theta}_N \, 1^N \right\|^2$ between the data point and its projection is constant—independent of θ. Thus, the Euclidean distance squared $\left\| y^N - \theta \, 1^N \right\|^2$ between the data and model points is, *up to an additive constant*, given by the Euclidean distance squared

$$\left\| \hat{\theta}_N \, 1^N - \theta \, 1^N \right\|^2 = \left(\hat{\theta}_N - \theta \right)^2 \left\| 1^N \right\|^2$$

between the projection and the model point. Hence, all we need to restore the Euclidean distance squared $\left\| y^N - \hat{\theta}_N \, 1^N \right\|^2$ with precision up to an additive constant is to know the estimate $\hat{\theta}_N$ and the Euclidean norm of 1^N. With respect to (3.2), this knowledge is equivalent to storing (and recursively updating if necessary) two values

$$\left(y^N \right) \left(1^N \right)^T = \sum_{k=1}^N y_k,$$

$$\left(1^N \right) \left(1^N \right)^T = \sum_{k=1}^N 1 = N.$$

Linear Regression

Orthogonal Projection. The least-squares matching of the sequence of observed data $(y_2, \ldots, y_{N+1})$ with a scalar θ-multiple of the time-lagged data sequence $(y_1, \ldots, y_N)$ according to the model

$$y_k = \theta y_{k-1} + e_k, \quad k = 1, \ldots, N$$

minimizes the Euclidean distance squared between both the vectors

$$\min_\theta \| y_2^{N+1} - \theta y_1^N \|^2.$$

We denote the solution to the problem by $\hat{\theta}_N$ to stress that the estimate is based on N data points

$$(y_2, y_1), \ldots, (y_{N+1}, y_N).$$

A necessary condition for the optimum $\hat{\theta}_N$ follows by differentiating with respect to θ

$$\frac{d}{d\theta} \| y_2^{N+1} - \hat{\theta}_N y_1^N \|^2 = -2 \left(y_2^{N+1} - \hat{\theta}_N y_1^N \right) \left(y_1^N \right)^T = 0.$$

This implies the condition

$$\left(y_2^{N+1} - \hat{\theta}_N y_1^N \right) \left(y_1^N \right)^T = 0 \tag{3.4}$$

which can be read so that the vector $y_2^{N+1} - \hat{\theta}_N y_1^N$ is orthogonal to the basis vector y_1^N (cf. Fig. 3.2). To put it a different way, $\hat{\theta}_N y_1^N$ is the orthogonal projection of y_2^{N+1} onto the straight line $\{ \theta y_1^N : \theta \in \mathbb{R} \}$. The solution to (3.4) is

$$\hat{\theta}_N = \frac{\left(y_2^{N+1} \right) \left(y_1^N \right)^T}{\left(y_1^N \right) \left(y_1^N \right)^T} = \frac{\sum_{k=2}^{N+1} y_k y_{k-1}}{\sum_{k=2}^{N+1} y_{k-1}^2} \tag{3.5}$$

provided that $\| y_1^N \| > 0$.

Pythagorean Relationship. To prove that $\hat{\theta}_N$ is even the global minimum, we first rewrite

$$\begin{aligned} \| y_2^{N+1} - \theta y_1^N \|^2 &= \| y_2^{N+1} - \hat{\theta}_N y_1^N + \hat{\theta}_N y_1^N - \theta y_1^N \|^2 \\ &= \| y_2^{N+1} - \hat{\theta}_N y_1^N \|^2 + \| \hat{\theta}_N y_1^N - \theta y_1^N \|^2 \\ &\quad + 2 \left(y_2^{N+1} - \hat{\theta}_N y_1^N \right)^T \left(\hat{\theta}_N y_1^N - \theta y_1^N \right). \end{aligned}$$

Taking into account (3.4), we see that the orthogonal projection $\hat{\theta}_N y_1^N$ satisfies the *Pythagorean relationship*

$$\| y_2^{N+1} - \theta y_1^N \|^2 = \| y_2^{N+1} - \hat{\theta}_N y_1^N \|^2 + \| \hat{\theta}_N y_1^N - \theta y_1^N \|^2. \tag{3.6}$$

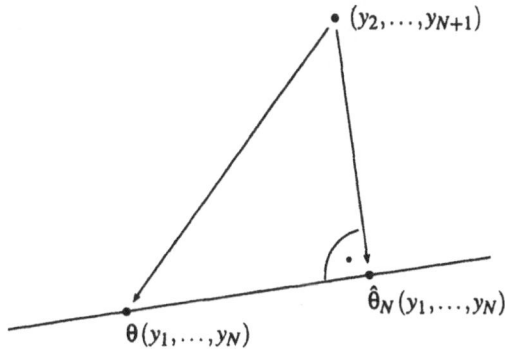

Fig. 3.2. The point $\hat{\theta}_N(y_1,\ldots,y_N)$ minimizing the Euclidean distance between the point $(y_2,\ldots,y_{N+1})$ and the straight line $\{\theta(y_1,\ldots,y_N):\theta \in \mathbb{R}\}$ is the orthogonal projection of y_2^{N+1} onto the straight line.

Hence, the Euclidean distance squared between y_2^{N+1} and any point θy_1^N is equal to a sum of the Euclidean distance squared between y_2^{N+1} and the orthogonal projection $\hat{\theta}_N y_1^N$ and the Euclidean distance squared between $\hat{\theta}_N y_1^N$ and θy_1^N. The *global minimum* property easily follows from the Pythagorean relationship: since for every θ

$$\left\| \hat{\theta}_N y_1^N - \theta y_1^N \right\|^2 \geq 0,$$

we have

$$\left\| y_2^{N+1} - \theta y_1^N \right\|^2 \geq \left\| y_2^{N+1} - \hat{\theta}_N y_1^N \right\|^2$$

with equality if and only if $\theta = \hat{\theta}_N$ provided $\left\| y_1^N \right\| > 0$.

Uniqueness of Projection. The uniqueness of the orthogonal projection easily follows. Suppose that two different points $\hat{\theta} y_1^N$ and $\hat{\theta}' y_1^N$, $\hat{\theta} \neq \hat{\theta}'$ satisfy (3.4). By Pythagorean relationship (3.6) it must hold

$$\left\| y_2^{N+1} - \hat{\theta} y_1^N \right\|^2 \geq \left\| y_2^{N+1} - \hat{\theta}' y_1^N \right\|^2$$

and, at the same time,

$$\left\| y_2^{N+1} - \hat{\theta}' y_1^N \right\|^2 \geq \left\| y_2^{N+1} - \hat{\theta} y_1^N \right\|^2.$$

Consequently, it must be

$$\left\| y_2^{N+1} - \hat{\theta}' y_1^N \right\|^2 = \left\| y_2^{N+1} - \hat{\theta} y_1^N \right\|^2$$

but the equality appears if and only if $\hat{\theta}' = \hat{\theta}$ which is contradiction with the assumption. Therefore, the orthogonal projection is unique.

Data Compression. Since the Euclidean distance squared $\left\|y_2^{N+1} - \hat{\theta}_N y_1^N\right\|^2$ between the data point and its projection is constant—independent of θ, the Euclidean distance squared $\left\|y_2^{N+1} - \theta y_1^N\right\|^2$ between the data and model points is, *up to an additive constant*, given by the Euclidean distance squared

$$\left\|\hat{\theta}_N y_1^N - \theta y_1^N\right\|^2 = \left(\hat{\theta}_N - \theta\right)^2 \left\|y_1^N\right\|^2$$

between the projection and the model point. Thus, all we need to reconstruct the Euclidean distance squared $\left\|y_2^{N+1} - \hat{\theta}_N y_1^N\right\|^2$ with precision up to an additive constant is to know the estimate $\hat{\theta}_N$ and the Euclidean norm y_1^N. With respect to (3.5), this knowledge is equivalent to storing (and possibly recursively updating) the following values

$$\left(y_2^{N+1}\right)\left(y_1^N\right)^T = \sum_{k=2}^{N+1} y_k y_{k-1},$$

$$\left(y_1^N\right)\left(y_1^N\right)^T = \sum_{k=2}^{N+1} y_{k-1}^2.$$

General Linear Regression.

Consider a general linear-in-parameter regression-type model

$$y_{m+1}^{N+m} = o + \theta^T H + e_{m+1}^{N+m}$$

where $\theta \in \mathbb{R}^n$ is a column vector of regression coefficients, o is a fixed row vector of dimension N and H is a matrix of type (n, N) whose rows are formed by basis vectors h_i, $i = 1, \ldots, n$, i.e., fixed row vectors chosen beforehand from $\mathbb{R}^N$

$$H = \begin{bmatrix} h_1 \\ h_2 \\ \vdots \\ h_n \end{bmatrix}.$$

Let us stress that H may depend on the observed data.

Orthogonal Projection. The least-squares matching of the sequence of observed data $y_{m+1}^{N+m} = (y_{m+1}, \ldots, y_{N+m})$ with $o + \theta^T H$ minimizes the Euclidean distance squared between both the vectors

$$\min_{\theta} \left\|y_{m+1}^{N+m} - o - \theta^T H\right\|^2.$$

A necessary condition for the optimum $\hat{\theta}_N$ follows by taking gradient with respect to θ

$$\nabla_\theta \left\|y_{m+1}^{N+m} - o - \hat{\theta}_N^T H\right\|^2 = -2\left(y_{m+1}^{N+m} - o - \hat{\theta}_N^T H\right) H^T = 0.$$

This implies the condition

$$\boxed{\left(y_{m+1}^{N+m} - o - \hat{\theta}_N^T H\right) H^T = 0}$$ (3.7)

which means that the vector $y_{m+1}^{N+m} - o - \hat{\theta}_N^T H$ is orthogonal to all rows in H (cf. Fig. 3.3). In other words, $o + \hat{\theta}_N^T H$ is an orthogonal projection of y_{m+1}^{N+m} onto the hyperplane $\{o + \theta^T H : \theta \in \mathbb{R}^n\}$. One can easily compute from

$$\left(y_{m+1}^{N+m} - o\right) H^T = \hat{\theta}_N^T H H^T$$

that the solution to (3.7) is

$$\hat{\theta}_N = \left(H H^T\right)^{-1} H (y_{m+1}^{N+m} - o)^T$$

provided that $H H^T$ is positive definite.

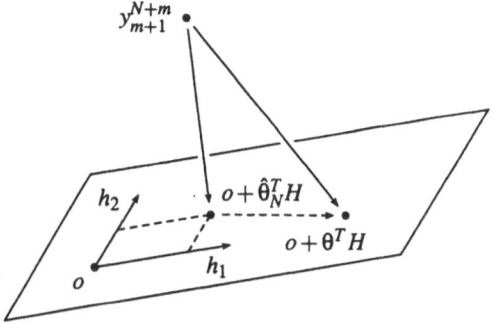

Fig. 3.3. The point $o + \hat{\theta}_N^T H$ minimizing the Euclidean distance between the point y_{m+1}^{N+m} and the hyperplane $\{o + \theta^T H : \theta \in \mathbb{R}^n\}$ is the orthogonal projection of y_{m+1}^{N+m} onto the hyperplane.

Pythagorean Relationship. To show that $\hat{\theta}_N$ is the global minimum, we first rewrite

$$\begin{aligned}
\left\|y_{m+1}^{N+m} - o - \theta^T H\right\|^2 &= \left\|y_{m+1}^{N+m} - o - \hat{\theta}_N^T H + \hat{\theta}_N^T H - \theta^T H\right\|^2 \\
&= \left\|y_{m+1}^{N+m} - o - \hat{\theta}_N^T H\right\|^2 + \left\|\hat{\theta}_N^T H - \theta^T H\right\|^2 \\
&\quad + 2 \left(y_{m+1}^{N+m} - o - \hat{\theta}_N^T H\right) \left(\hat{\theta}_N^T H - \theta^T H\right)^T.
\end{aligned}$$

With respect to (3.7), the orthogonal projection $\hat{\theta}_N y_1^N$ satisfies the *Pythagorean relationship*

$$\boxed{\left\|y_{m+1}^{N+m} - o - \theta^T H\right\|^2 = \left\|y_{m+1}^{N+m} - o - \hat{\theta}_N^T H\right\|^2 + \left\|\hat{\theta}_N^T H - \theta^T H\right\|^2.}$$ (3.8)

Hence, the Euclidean distance squared between y_{m+1}^{N+m} and any point $o + \theta^T H$ is equal to a sum of the Euclidean distance squared between y_{m+1}^{N+m} and the orthogonal projection $o + \hat{\theta}_N^T H$ and the Euclidean distance squared between $o + \hat{\theta}_N^T H$ and $o + \theta^T H$.

The proof of the global minimum property and uniqueness of $\hat{\theta}_N$ is completely analogous to the special cases shown above.

3.2 Kullback-Leibler Distance

The role of Kullback-Leibler distance in probability matching is similar to the role of Euclidean distance in data matching. The present section sums up some of the fundamental properties of Kullback-Leibler distance we shall need in the following.

Definitions. Kullback-Leibler distance between two probability density functions $s(y)$ and $s'(y)$ is defined as

$$D(s\|s') \triangleq \int s(y) \log \frac{s(y)}{s'(y)} \, dy \tag{3.9}$$

where the logarithm is understood to the base e. Note that (3.9) is well defined only when for every (Borel) set $A \subset \mathcal{Y}$

$$\int_A s'(y) \, dy = 0 \quad \text{implies} \quad \int_A s(y) \, dy = 0.$$

In the opposite case, K-L distance of $s(y)$ and $s'(y)$ is set infinite, $D(s\|s') = \infty$.

Analogously, we can define *conditional Kullback-Leibler distance* between two conditional probability density functions $s(y|z)$ and $s'(y|z)$ given a marginal density $\tilde{r}(z) = \int r(y, z) \, dy$ as

$$\bar{D}(s\|s'|\tilde{r}) \triangleq \int\int s(y|z) \, \tilde{r}(z) \log \frac{s(y|z)}{s'(y|z)} \, dy \, dz. \tag{3.10}$$

Note that the definition (3.10) can be formally rewritten as

$$\bar{D}(s\|s'|\tilde{r}) = D(s\tilde{r}\|s'\tilde{r}) \tag{3.11}$$

using the convention $0\log \frac{0}{0} = 0$ whenever $\tilde{r}(z) = 0$.

Alternatively, we can define *conditional Kullback-Leibler distance* between the joint density $r(y, z)$ and conditional density $s(y|z)$ as

$$\bar{D}(r\|s) \triangleq \int\int r(y, z) \log \frac{r(y, z)}{s(y|z) \, \tilde{r}(z)} \, dy \, dz. \tag{3.12}$$

The definition (3.12) can be formally rewritten as

$$\bar{D}(r\|s) = D(r\|s\tilde{r}) \tag{3.13}$$

using again the convention $0\log \frac{0}{0} = 0$ whenever $\tilde{r}(z) = 0$.

Kullback-Leibler distance will be abbreviated as K-L distance in the sequel.

Relation to Inaccuracy. The inaccuracy $K(r:s)$ of $r(y)$ relative to $s(y)$ can be decomposed as follows

$$K(r\|s) = \int r(y) \log \frac{1}{s(y)} \, dy$$

$$= \int r(y) \log \frac{r(y)}{s(y)} \, dy + \int r(y) \log \frac{1}{r(y)} \, dy$$

provided all the integrals exist.

Introducing (differential) *Shannon's entropy* of $r(y)$ through

$$H(r) \triangleq \int r(y) \log \frac{1}{r(y)} \, dy \tag{3.14}$$

we obtain thus the identity

$$\boxed{K(r:s) = D(r\|s) + H(r)} \tag{3.15}$$

that shows how inaccuracy, K-L distance and Shannon's entropy are related to each other.

Nonnegativity of K-L Distance. In the sequel we often need the fact that K-L distance is nonnegative.

Let $\mathcal{Y}_s = \{y : s(y) > 0\}$ be the support set of $s(y)$. Since the function $f(u) = -\log u$ is convex on $(0, \infty)$, we can deduce by application of Jensen's inequality (B.4) that

$$D(s\|s') = \int_{\mathcal{Y}_s} s(y) \log \frac{s(y)}{s'(y)} \, dy$$

$$= \int_{\mathcal{Y}_s} s(y) \left(-\log \frac{s'(y)}{s(y)} \right) dy$$

$$\geq -\log \int_{\mathcal{Y}_s} s(y) \frac{s'(y)}{s(y)} \, dy$$

$$= -\log \int_{\mathcal{Y}_s} s'(y) \, dy$$

$$\geq -\log \int_{\mathcal{Y}} s'(y) \, dy$$

$$= -\log 1 = 0.$$

Since $f(u) = -\log u$ is strictly convex, the equality in the above relation appears if and only if $s'(y)/s(y) = 1$, i.e., $s(y) = s'(y)$ (almost everywhere on $\mathcal{Y}$, i.e., except perhaps sets of Lebesgue measure 0).

To sum up, K-L distance of two density functions s and s' is always nonnegative

$$\boxed{D(s\|s') \geq 0} \tag{3.16}$$

with equality if and only if $s(y) = s'(y)$ (almost everywhere on $\mathcal{Y}$).

An analogous property holds for the conditional K-L distance (3.10) as well. With the interpretation (3.11), it follows immediately by analogy with (3.16) that

$$\bar{D}(s\|s'|\bar{r}) \geq 0 \tag{3.17}$$

with equality if and only if $s(y|z)\,\bar{r}(z) = s'(y|z)\,\bar{r}(z)$ (almost everywhere on $\mathcal{Y} \times \mathcal{Z}$).

A similar result holds for the conditional K-L distance defined as (3.12). With the interpretation (3.13), it follows again by analogy with (3.16) that

$$\bar{D}(r\|s) \geq 0 \tag{3.18}$$

with equality if and only if $r(y,z) = s(y|z)\,\bar{r}(z)$ (almost everywhere on $\mathcal{Y} \times \mathcal{Z}$).

Asymmetry of K-L distance. The value of K-L distance $D(s\|s')$ clearly depends on the order of its arguments $s(y)$, $s'(y)$. K-L distance thus lacks one of the key properties of the true distance.

Example 3.1 (*Normal distributions*) Consider two normal density functions with the parameters $\theta_i = (\mu_i, \sigma_i^2)$

$$s_{\theta_1}(y) = \frac{1}{\sqrt{2\pi\sigma_1^2}} \exp\left(-\frac{1}{2\sigma_1^2}(y-\mu_1)^2\right),$$

$$s_{\theta_2}(y) = \frac{1}{\sqrt{2\pi\sigma_2^2}} \exp\left(-\frac{1}{2\sigma_2^2}(y-\mu_2)^2\right).$$

K-L distance of $s_{\theta_1}(y)$ and $s_{\theta_2}(y)$ is by definition (3.9)

$$
\begin{aligned}
D(s_{\theta_1}\|s_{\theta_2}) &= \int s_{\theta_1}(y) \log \frac{(2\pi\sigma_1^2)^{-\frac{1}{2}} \exp\left(-\frac{1}{2\sigma_1^2}(y-\mu_1)^2\right)}{(2\pi\sigma_2^2)^{-\frac{1}{2}} \exp\left(-\frac{1}{2\sigma_2^2}(y-\mu_2)^2\right)} \, dy \\
&= -\frac{1}{2}\log\frac{\sigma_1^2}{\sigma_2^2} - \frac{1}{2\sigma_1^2}\int s_{\theta_1}(y)(y-\mu_1)^2 \, dy + \frac{1}{2\sigma_2^2}\int s_{\theta_1}(y)(y-\mu_2)^2 \, dy \\
&= -\frac{1}{2}\log\frac{\sigma_1^2}{\sigma_2^2} - \frac{1}{2\sigma_1^2}\sigma_1^2 + \frac{1}{2\sigma_2^2}\left(\sigma_1^2 + (\mu_1-\mu_2)^2\right)
\end{aligned}
$$

so that we have

$$D(s_{\theta_1}\|s_{\theta_2}) = \frac{1}{2}\left(\frac{\sigma_1^2}{\sigma_2^2} - \log\frac{\sigma_1^2}{\sigma_2^2} - 1\right) + \frac{(\mu_1-\mu_2)^2}{2\sigma_2^2}. \tag{3.19}$$

If the variances are equal, $\sigma_1^2 = \sigma_2^2$, K-L distance is proportional to the square of the means $(\mu_1 - \mu_2)^2$ (cf. Fig. 3.4). In this special case, K-L distance does not depend on the order of arguments so that $D(s_{\theta_1}\|s_{\theta_2}) = D(s_{\theta_2}\|s_{\theta_1})$.

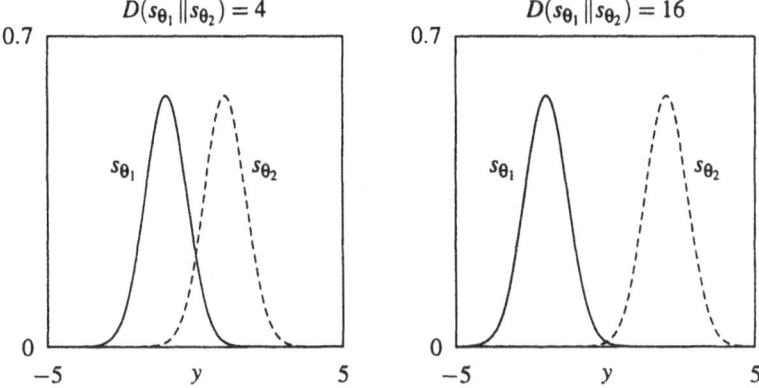

Fig.3.4. K-L distance of normal distributions with different means.

The symmetry of K-L distance disappears, however, as soon as the variances differ, $\sigma_1^2 \neq \sigma_2^2$. The value of K-L distance depends then on the order of arguments, i.e., by interchanging $s_{\theta_1}(y)$ and $s_{\theta_2}(y)$ in $D(s_{\theta_1} \| s_{\theta_2})$ we may obtain quite different values of K-L distance (see Fig. 3.5). The reason is that K-L distance is much more sensitive to the tail behaviour of the second argument $s_{\theta_2}(y)$. $\qquad\qquad\square$

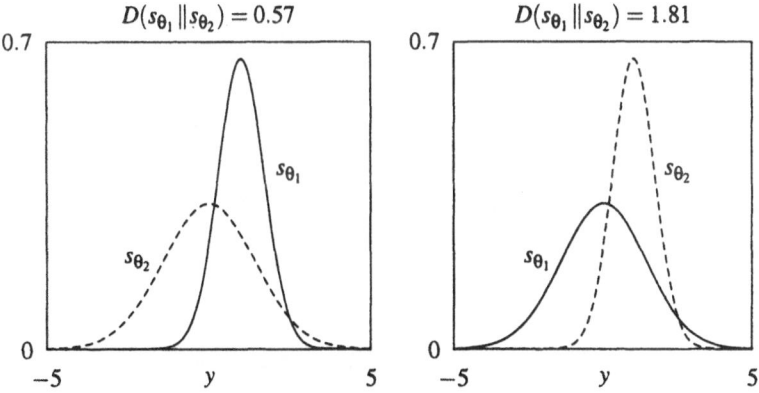

Fig.3.5. K-L distance of normal distributions with different variances.

Example 3.2 (*Distributions with different supports*) The asymmetry of K-L distance becomes really drastic when the density functions $s(y)$ and $s'(y)$ have different supports. Let us denote by

$$\mathcal{Y}_s = \{y : s(y) > 0\}, \quad \mathcal{Y}_{s'} = \{y : s'(y) > 0\}$$

the supports of $s(y)$ and $s'(y)$, respectively. If $\mathcal{Y}_{s'} \subset \mathcal{Y}_s$ but $\mathcal{Y}_s \not\subset \mathcal{Y}_{s'}$, then $D(s\|s') = \infty$ though $D(s'\|s)$ may still be finite (cf. Fig. 3.6).   □

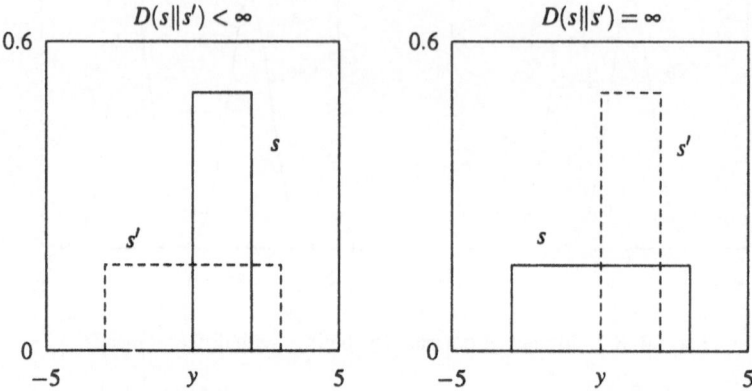

Fig. 3.6. K-L distance of distributions with different supports.

Convexity. K-L distance $D(s\|s')$ can be shown to be convex in both s and s'.

Since the function $f(t) = t \log t$ is convex on $(0, \infty)$ (check that $f''(t) = \frac{1}{t} > 0$ for all positive t), we have by Jensen's inequality (B.4)

$$f\big(\mu t_1 + (1-\mu) t_2\big) \le \mu f(t_1) + (1-\mu) f(t_2)$$

for any $0 \le \mu \le 1$. The substitution

$$t_1 = \frac{\lambda s_1(y)}{\lambda s'_1(y)}, \quad t_2 = \frac{(1-\lambda) s_2(y)}{(1-\lambda) s'_2(y)}, \quad \mu = \frac{\lambda s'_1(y)}{\lambda s'_1(y) + (1-\lambda) s'_2(y)}$$

gives

$$\big(\lambda s_1(y) + (1-\lambda) s_2(y)\big) \log \frac{\lambda s_1(y) + (1-\lambda) s_2(y)}{\lambda s'_1(y) + (1-\lambda) s'_2(y)}$$

$$\le \lambda s_1(y) \log \frac{s_1(y)}{s'_1(y)} + (1-\lambda) s_2(y) \log \frac{s_2(y)}{s'_2(y)}.$$

After integrating both sides of the inequality, we obtain

$$\boxed{D\big(\lambda s_1 + (1-\lambda) s_2 \,\|\, \lambda s'_1 + (1-\lambda) s'_2\big) \le \lambda D(s_1\|s'_1) + (1-\lambda) D(s_2\|s'_2)} \quad (3.20)$$

for all $0 \le \lambda \le 1$. In other words, K-L distance $D(r\|s)$ is convex in the pair (r, s).

Example 3.3 *(Probability vectors)* Consider two probability mass functions $[a, 1-a]$ and $[b, 1-b]$ on a two-point set $\{1, 2\}$. Their K-L distance is, by direct analogy with (3.9),

$$D([a, 1-a]\|[b, 1-b]) = a \log \frac{a}{b} + (1-a) \log \frac{1-a}{1-b}.$$

Fixing $[a, 1-a] = [0.3, 0.7]$, K-L distance is a function of the scalar $b \in (0,1)$. The dependence is plotted in Fig. 3.7. Note that K-L distance reaches its minimum—zero value at $b = a$. □

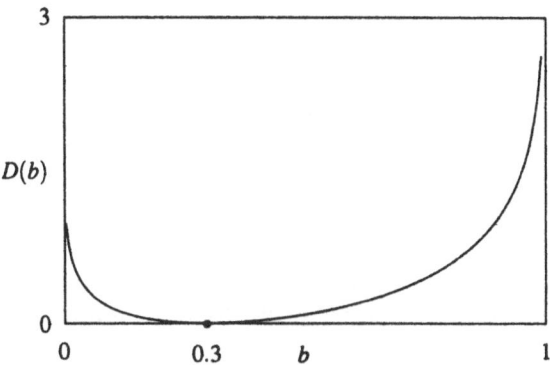

Fig. 3.7. Kullback-Leibler distance between two mass functions $[a, 1-a]$ and $[b, 1-b]$ plotted as a function of b given $a = 0.3$.

Example 3.4 (*Probability vectors*)   Consider two probability mass functions $[a_1, a_2, a_3]$ and $[b_1, b_2, b_3]$ on a three-point set $\{1, 2, 3\}$. Setting $[a_1, a_2, a_3] = [1/6, 2/6, 3/6]$, the K-L distance $D([a_1, a_2, a_3]\|[b_1, b_2, b_3])$ is a function of $b_1 \in (0,1)$, $b_2 \in (0, 1-b_1)$. This dependence is envisaged by a contour graph in Fig. 3.8. The minimum—zero value of K-L distance is obtained for $b_1 = a_1$, $b_2 = a_2$. □

Locally Euclidean Behaviour.   Consider two parameter vectors θ and θ' such that the difference $\Delta\theta = \theta' - \theta$ is sufficiently small. Performing Taylor's expansion of $\log s_{\theta'}(y)$ around θ and neglecting higher than second-order terms, we obtain

$$\log s_{\theta'}(y) \approx \log s_\theta(y) + (\theta' - \theta)^T \nabla_\theta \log s_\theta(y) + \frac{1}{2}(\theta' - \theta)^T \nabla_\theta^2 \log s_\theta(y)(\theta' - \theta)$$

Taking expectation of $\log s_{\theta'}(Y) - \log s_\theta(Y)$ with respect to s_θ, we have

$$\int s_\theta(y) \log \frac{s_\theta(y)}{s_{\theta'}(y)} dy$$
$$\approx -(\theta' - \theta)^T \left(\int s_\theta(y) \nabla_\theta \log s_\theta(y) dy \right)$$
$$-\frac{1}{2}(\theta' - \theta)^T \left(\int s_\theta(y) \nabla_\theta^2 \log s_\theta(y) dy \right)(\theta' - \theta) \qquad (3.21)$$

where

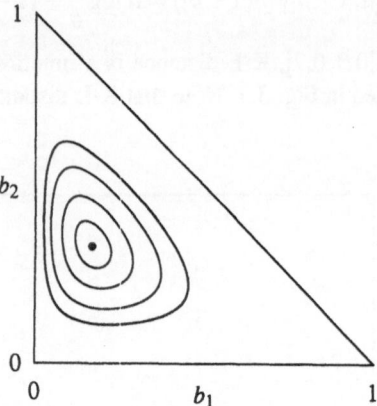

Fig.3.8. Curves of equal Kullback-Leibler distance defined on a 2-simplex by the equation $D([a_1,a_2,a_3]\|[b_1,b_2,b_3]) = d$ with a fixed distribution $[a_1,a_2,a_3] = [1/6,2/6,3/6]$ and $d = 0, 0.01, 0.04, 0.1, 0.18$.

$$(\nabla_\theta \log s_\theta(y))_i = \frac{\partial}{\partial \theta_i} \log s_\theta(y)$$

$$(\nabla_\theta^2 \log s_\theta(y))_{i,j} = \frac{\partial^2}{\partial \theta_i \partial \theta_j} \log s_\theta(y)$$

denote a column vector of first-order derivatives and matrix of second-order derivatives of $\log s_\theta(y)$ with respect to particular entries of the vector θ.

The first term is zero because

$$\int s_\theta(y) \nabla_\theta \log s_\theta(y)\, dy = \int s_\theta(y) \frac{\nabla_\theta s_\theta(y)}{s_\theta(y)}\, dy$$

$$= \int \nabla_\theta s_\theta(y)\, dy$$

$$= \nabla_\theta \int s_\theta(y)\, dy$$

$$= \nabla_\theta 1 = 0. \tag{3.22}$$

The second term can be rewritten as follows

$$\int s_\theta(y) \nabla_\theta^2 \log s_\theta(y)\, dy$$

$$= \int s_\theta(y) \frac{(\nabla_\theta^2 s_\theta(y)) s_\theta(y) - (\nabla_\theta s_\theta(y))(\nabla_\theta s_\theta(y))^T}{(s_\theta(y))^2}\, dy$$

$$= \int \nabla_\theta^2 s_\theta(y)\, dy - \int s_\theta(y) \left(\frac{\nabla_\theta s_\theta(y)}{s_\theta(y)}\right)\left(\frac{\nabla_\theta s_\theta(y)}{s_\theta(y)}\right)^T dy$$

$$= \nabla_\theta^2 \int s_\theta(y)\, dy - \int s_\theta(y) (\nabla_\theta \log s_\theta(y))(\nabla_\theta \log s_\theta(y))^T dy$$

$$= -\int s_\theta(y) \left(\nabla_\theta \log s_\theta(y)\right) \left(\nabla_\theta \log s_\theta(y)\right)^T dy. \tag{3.23}$$

Introducing the *Fisher information matrix*

$$I(\theta) \triangleq \int s_\theta(y) \left(\nabla_\theta \log s_\theta(y)\right) \left(\nabla_\theta \log s_\theta(y)\right)^T dy$$

and substituting (3.22) and (3.23) into (3.21), we find that K-L distance of two neighbouring densities s_θ and $s_{\theta'}$ for θ' near θ can be approximated—neglecting higher than second-order terms—by the quadratic form

$$\boxed{D(s_\theta \| s_{\theta'}) \approx \frac{1}{2} (\theta' - \theta)^T I(\theta) (\theta' - \theta).} \tag{3.24}$$

3.3 Pythagorean View of Statistical Estimation

Using the above properties of K-L distance, we develop now a geometric picture of probability-based estimation, similar to the Euclidean picture of data matching given above. The fundamental tool used here is a *Pythagorean relationship* that enables us to decompose inaccuracy into a sum of two terms, one independent of the unknown parameter, the other independent of data given a certain statistic.

There is a substantial difference between the cases of independent observations and controlled dynamic systems. Since in the latter case we deal with conditional densities, the Pythagorean relationship needs essentially to be applied *twice*.

Independent Observations

Assumptions. We consider a family of density functions

$$\mathcal{S} \triangleq \left\{ s_\theta(y) : \theta \in \mathcal{T} \right\}$$

that satisfies the following *regularity conditions*

(a) $\mathcal{T}$ is an open set in $\mathbb{R}^{\dim \theta}$,

(b) $s_\theta(y) > 0$ for all $y \in \mathcal{Y}$ and all $\theta \in \mathcal{T}$,

(c) $s_\theta(y)$ is continuously differentiable in y for all $\theta \in \mathcal{T}$.

We show that if $\mathcal{S}$ can be imbedded in an n-dimensional exponential family then it is sufficient to know the minimum inaccuracy projection of the empirical density $r_N(y)$ onto the exponential family in order to restore the inaccuracy $K(r_N : s_\theta)$ with precision up to an additive constant.

Exponential Family. Let $s_\theta(y)$ be an arbitrary fixed density from $\mathcal{S}$. We construct an exponential family $\mathcal{S}_{\theta;h}$ composed of densities

$$s_{\theta,\lambda}(y) = s_\theta(y) \exp\big(\lambda^T h(y) - \psi(\theta,\lambda)\big) \tag{3.25}$$

where $\lambda \in \mathbb{R}^n$ is a vector parameter of the family, the function $h: \mathcal{Y} \to \mathbb{R}^n$ is a given vector function (statistic) of single observation y and

$$\psi(\theta,\lambda) = \log \int s_\theta(y) \exp\big(\lambda^T h(y)\big) \, dy \tag{3.26}$$

is logarithm of the normalizing divisor.

Note that λ is usually called the natural or canonical parameter, h is called the canonical statistic and ψ is also known as the cumulant generating function.

Further on it is assumed that the functions $h_0(y) \equiv 1, h_1(y), \ldots, h_n(y)$ are linearly independent. Since two densities $s_{\theta,\lambda}(y)$ and $s_{\theta,\lambda'}(y)$ are equal if and only if the right-hand side of

$$\log \frac{s_{\theta,\lambda}(y)}{s_{\theta,\lambda'}(y)} = (\lambda - \lambda')^T h(y) - \psi(\theta,\lambda) + \psi(\theta,\lambda')$$

vanishes, the assumption implies a one-to-one correspondence between the vector parameter λ and the density $s_{\theta,\lambda}(y)$. The dimension of $\mathcal{S}_{\theta;h}$ then equals n.

Normalizing Divisor. The parameter λ is assumed to run through all values from $\mathbb{R}^n$ for which the normalizing divisor is finite

$$\exp\big(\psi(\theta,\lambda)\big) < \infty.$$

We denote the set of all such λ's by $\mathcal{N}_\theta$. It is easy to verify that the set $\mathcal{N}_\theta$ is convex. Let λ and λ' be two parameter points for which the normalizing divisor is finite. Then by Hölder inequality (A.31) for every $0 < a < 1$

$$\int s_\theta(y) \exp\Big((a\lambda + (1-a)\lambda')^T h(y)\Big) \, dy$$
$$\leq \left(\int s_\theta(y) \exp(\lambda^T h(y)) \, dy\right)^a \left(\int s_\theta(y) \exp(\lambda'^T h(y)) \, dy\right)^{1-a} < \infty.$$

By taking logarithms of both sides, we obtain

$$\log \int s_\theta(y) \exp\Big((a\lambda + (1-a)\lambda')^T h(y)\Big) \, dy$$
$$\leq a \log\left(\int s_\theta(y) \exp(\lambda^T h(y)) \, dy\right)$$
$$+ (1-a) \log\left(\int s_\theta(y) \exp(\lambda'^T h(y)) \, dy\right).$$

Hence, the function $\psi(\theta,\lambda)$ is a convex function of λ on the convex set $\mathcal{N}_\theta$.

In the sequel, $\mathcal{S}_{\theta;h}$ is understood to be the maximal family of densities that can be expressed in the form (3.25) for some $\lambda \in \mathbb{R}^n$.

h-Projection. Suppose that a sample y^N is given with the empirical density $r_N(y)$. The necessary condition for $\hat{\lambda}$ to be the minimum inaccuracy (maximum likelihood) estimate of the parameter λ is

$$0 = \nabla_\lambda K(r_N : s_{\theta,\hat{\lambda}})$$

$$= \int r_N(y) \left(-h(y) + \int s_{\theta,\hat{\lambda}}(y) h(y) \, dy \right) dy$$

$$= -\int r_N(y) h(y) \, dy + \int s_{\theta,\hat{\lambda}}(y) h(y) \, dy.$$

The condition thus reads

$$\boxed{\int s_{\theta,\hat{\lambda}}(y) h(y) \, dy = \int r_N(y) h(y) \, dy.} \tag{3.27}$$

Note that the estimate $\hat{\lambda}$ depends in general on N and θ but for simplicity we do not stress the fact in our notation.

The density $s_{\theta,\hat{\lambda}}(y)$ that satisfies the condition (3.27) is called a *h-projection* of $r_N(y)$ onto $S_{\theta;h}$. Introducing the notation

$$\bar{h}_N \triangleq \int r_N(y) h(y) \, dy = \frac{1}{N} \sum_{k=1}^{N} h(y_k), \tag{3.28}$$

$$\hat{h}(\theta,\lambda) \triangleq \int s_{\theta,\lambda}(y) h(y) \, dy, \tag{3.29}$$

we can write (3.27) as

$$\hat{h}(\theta,\hat{\lambda}) = \bar{h}_N.$$

We denote the set of all densities $r(y)$ with the same h-projection as

$$\mathcal{R}_N \triangleq \left\{ r(y) : \int r(y) h(y) \, dy = \bar{h}_N, \int r(y) \, dy = 1, \, r(y) \geq 0 \right\}.$$

The set depends on the canonical statistic $h(y)$ but for simplicity again, we do not stress it in our notation.

The expectation $\hat{h}(\theta,\lambda)$ can be regarded as an alternative way of parametrizing the exponential family $S_{\theta;h}$, dual in a sense to the canonical λ-parametrization. The close link between both is done by the fact that $\hat{h}(\theta,\lambda)$ coincides with the gradient of logarithm of the normalizing divisor $\psi(\theta,\lambda)$ with respect to λ

$$\nabla_\lambda \psi(\theta,\lambda) = \nabla_\lambda \log \int s_\theta(y) \exp(\lambda^T h(y)) \, dy$$

$$= \frac{\int s_\theta(y) \nabla_\lambda \exp(\lambda^T h(y)) \, dy}{\int s_\theta(y) \exp(\lambda^T h(y)) \, dy}$$

$$= \int \frac{s_\theta(y) \exp(\lambda^T h(y))}{\int s_\theta(y) \exp(\lambda^T h(y)) \, dy} h(y) \, dy$$

$$= \int s_{\theta,\lambda}(y) h(y) \, dy$$

$$= \hat{h}(\theta,\lambda). \tag{3.30}$$

Pythagorean Relationship. Let $s_{\theta,\lambda}(y)$ be exponential (3.25) and $\hat{\lambda}$ satisfy (3.27). Then we can write

$$K(r_N{:}s_\theta) - K(r_N{:}s_{\theta,\hat{\lambda}})$$

$$= \int r_N(y) \log \frac{s_{\theta,\hat{\lambda}}(y)}{s_\theta(y)} \, dy$$

$$= \int r_N(y) \log \frac{s_\theta(y) \exp(\hat{\lambda}^T h(y) - \psi(\theta,\hat{\lambda}))}{s_\theta(y)} \, dy$$

$$= \hat{\lambda}^T \left(\int r_N(y) h(y) \, dy \right) - \psi(\theta,\hat{\lambda})$$

$$= \hat{\lambda}^T \left(\int s_{\theta,\hat{\lambda}}(y) h(y) \, dy \right) - \psi(\theta,\hat{\lambda})$$

$$= \int s_{\theta,\hat{\lambda}}(y) \log \frac{s_\theta(y) \exp(\hat{\lambda}^T h(y) - \psi(\theta,\hat{\lambda}))}{s_\theta(y)} \, dy$$

$$= \int s_{\theta,\hat{\lambda}}(y) \log \frac{s_{\theta,\hat{\lambda}}(y)}{s_\theta(y)} \, dy$$

$$= D(s_{\theta,\hat{\lambda}} \| s_\theta).$$

As a result, we have

$$\boxed{K(r_N{:}s_\theta) = K(r_N{:}s_{\theta,\hat{\lambda}}) + D(s_{\theta,\hat{\lambda}} \| s_\theta)} \tag{3.31}$$

which can be regarded as an analogue of the Pythagorean relationship (cf. Fig. 3.9).

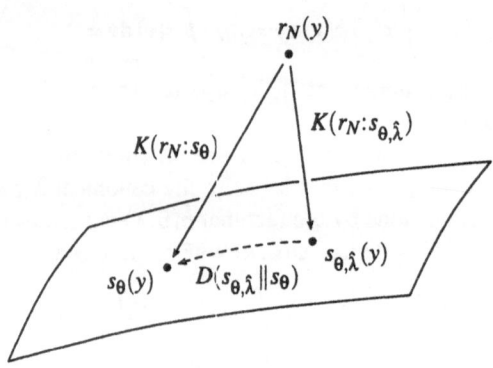

Fig. 3.9. Pythagorean-like decomposition of inaccuracy for independent observations.

Minimum Inaccuracy Projection. Assume that the inaccuracy $K(r_N{:}s_{\theta,\hat{\lambda}})$ of $r_N(y)$ relative to the h-projection $s_{\theta,\hat{\lambda}}(y)$ is finite

$$K(r_N:s_{\theta,\hat{\lambda}}) < \infty. \tag{3.32}$$

The following Pythagorean relationship holds for every $\lambda \in \mathcal{N}_\theta$

$$K(r_N:s_{\theta,\lambda}) - K(r_N:s_{\theta,\hat{\lambda}})$$

$$= \int r_N(y) \log \frac{s_{\theta,\hat{\lambda}}(y)}{s_{\theta,\lambda}(y)} \, dy$$

$$= \int r_N(y) \log \frac{s_\theta(y) \exp(\hat{\lambda}^T h(y) - \psi(\theta,\hat{\lambda}))}{s_\theta(y) \exp(\lambda^T h(y) - \psi(\theta,\lambda))} \, dy$$

$$= (\hat{\lambda} - \lambda)^T \left(\int r_N(y) h(y) \, dy \right) - \psi(\theta,\hat{\lambda}) + \psi(\theta,\lambda)$$

$$= (\hat{\lambda} - \lambda)^T \left(\int s_{\theta,\hat{\lambda}}(y) h(y) \, dy \right) - \psi(\theta,\hat{\lambda}) + \psi(\theta,\lambda)$$

$$= \int s_{\theta,\hat{\lambda}}(y) \log \frac{s_\theta(y) \exp(\hat{\lambda}^T h(y) - \psi(\theta,\hat{\lambda}))}{s_\theta(y) \exp(\lambda^T h(y) - \psi(\theta,\lambda))} \, dy$$

$$= \int s_{\theta,\hat{\lambda}}(y) \log \frac{s_{\theta,\hat{\lambda}}(y)}{s_{\theta,\lambda}(y)} \, dy$$

$$= D(s_{\theta,\hat{\lambda}} \| s_{\theta,\lambda}).$$

Since the K-L distance $D(s_{\theta,\hat{\lambda}} \| s_{\theta,\lambda})$ is nonnegative by (3.16) and the inaccuracy $K(r_N:s_{\theta,\hat{\lambda}})$ was supposed finite in (3.32), we have

$$K(r_N:s_{\theta,\lambda}) \geq K(r_N:s_{\theta,\hat{\lambda}})$$

with equality if and only if $s_{\theta,\lambda}(y) = s_{\theta,\hat{\lambda}}(y)$ (almost everywhere). Thus, under the assumption (3.32), the h-projection $s_{\theta,\hat{\lambda}}(y)$ is a unique solution to the minimum inaccuracy problem

$$\boxed{K(r_N:s_{\theta,\hat{\lambda}}) = \min_{\lambda \in \mathcal{N}_\theta} K(r_N:s_{\theta,\lambda}).} \tag{3.33}$$

Minimum K-L Distance Projection. The h-projection can be given another meaning, dual in a sense to the above one. Assume that K-L distance of the h-projection $s_{\theta,\hat{\lambda}}(y)$ and the model density $s_\theta(y)$ is finite

$$D(s_{\theta,\hat{\lambda}} \| s_\theta) < \infty. \tag{3.34}$$

Then there are $r(y) \in \mathcal{R}_N$ such that $D(r \| s_\theta) < \infty$. For every such $r(y) \in \mathcal{R}_N$, the following Pythagorean relation holds

$$D(r \| s_\theta) - D(r \| s_{\theta,\hat{\lambda}})$$

$$= \int r(y) \log \frac{s_{\theta,\hat{\lambda}}(y)}{s_\theta(y)} \, dy$$

$$
\begin{aligned}
&= \int r(y) \log \frac{s_\theta(y) \exp\left(\hat{\lambda}^T h(y) - \psi(\theta, \hat{\lambda})\right)}{s_\theta(y)} \, dy \\
&= \hat{\lambda}^T \left(\int r(y) h(y) \, dy \right) - \psi(\theta, \hat{\lambda}) \\
&= \hat{\lambda}^T \left(\int s_{\theta,\hat{\lambda}}(y) h(y) \, dy \right) - \psi(\theta, \hat{\lambda}) \\
&= \int s_{\theta,\hat{\lambda}}(y) \log \frac{s_\theta(y) \exp\left(\hat{\lambda}^T h(y) - \psi(\theta, \hat{\lambda})\right)}{s_\theta(y)} \, dy \\
&= \int s_{\theta,\hat{\lambda}}(y) \log \frac{s_{\theta,\hat{\lambda}}(y)}{s_\theta(y)} \, dy \\
&= D(s_{\theta,\hat{\lambda}} \| s_\theta).
\end{aligned}
$$

Since the K-L distance $D(r \| s_{\theta,\hat{\lambda}})$ is nonnegative by (3.16) and we consider only $r(y) \in \mathcal{R}_N$ such that $D(r \| s_\theta) < \infty$, we have for every such $r(y)$

$$
D(r \| s_\theta) \geq D(s_{\theta,\hat{\lambda}} \| s_\theta)
$$

with equality if and only if $r(y) = s_{\theta,\hat{\lambda}}(y)$ almost everywhere. Thus, under the assumption (3.34), the h-projection $s_{\theta,\hat{\lambda}}(y)$ is a unique solution to the minimum K-L distance problem

$$
\boxed{D(s_{\theta,\hat{\lambda}} \| s_\theta) = \min_{r \in \mathcal{R}_N} D(r \| s_\theta).} \tag{3.35}
$$

Note that the minimum K-L distance solution can be seen as generalization of the maximum entropy solution; in contrast to the maximum entropy principle, K-L distance is taken *relative* to the model density $s_\theta(y)$.

Orthogonal Projection. The Pythagorean relationship (3.31) can be rewritten in the following appealing form

$$
\int \left(r_N(y) - s_{\theta,\hat{\lambda}}(y) \right) \left(\log s_\theta(y) - \log s_{\theta,\hat{\lambda}}(y) \right) dy = 0. \tag{3.36}
$$

The above condition can be taken as a definition of *orthogonal projection* of r_N onto $\mathcal{S}_{\theta;h}$ or, from the dual viewpoint, as a definition of *orthogonal projection* of s_θ onto $\mathcal{R}_N$.

Compare the condition (3.36) with the orthogonal projection in the Euclidean sense for the model $y = \theta + e$, for instance,

$$
\left(y^N - \hat{\theta} 1^N \right) \left(\theta 1^N - \hat{\theta} 1^N \right) = 0.
$$

In contrast to what we are used to in the Euclidean space, the appearance of logarithm in (3.36) makes the condition of "orthogonality" asymmetric in $r(y)$ and $s(y)$. Consequently, in the space of probability density functions there is no straightforward analogy of the "natural" inner product as we know it in the Euclidean space.

To give a better insight into (3.36), we can rewrite it in a more symmetric way as

$$\int s_{\theta,\hat{\lambda}}(y) \frac{\partial}{\partial \mu} \log\left[\mu r_N(y) + (1-\mu) s_{\theta,\hat{\lambda}}(y)\right]\Big|_{\mu=0}$$

$$\cdot \frac{\partial}{\partial \lambda} \log\left[s_\theta(y) \exp(\lambda^T h(y) - \psi(\theta,\lambda))\right]\Big|_{\lambda=\hat{\lambda}} dy = 0.$$

This definition gives rise to a specific kind of Riemannian geometry on a differentiable manifold of probability density functions. The underlying metric tensor is closely related to the Fisher information matrix. In contrast to the classical Riemannian-geometric picture, two dual affine connections need to be considered at the same time to explain the above indicated asymmetry of the geometry. In these connections, exponential and mixture families of probability distributions act as analogy of hyperplanes in the Euclidean case.

We do not follow this view here. First, the systematic use of differential geometry would require mathematical tools going far beyond the level of this book. Second, the differential-geometric view of estimation can be developed rigorously only for certain special cases, e.g., when the observed data are discrete or when the model family is imbedded in a finite-dimensional exponential family. The reader interested in application of differential geometry in Bayesian estimation is referred to Kulhavý (1990; 1992; 1993; 1994b).

Minimum K-L Distance. One way of evaluating the K-L distance $D(s_{\theta,\hat{\lambda}} \| s_\theta)$ is to determine the h-projection $s_{\theta,\hat{\lambda}}(y)$ explicitly and substitute it in $D(s_{\theta,\hat{\lambda}} \| s_\theta)$. This results in

$$\begin{aligned}
D(s_{\theta,\hat{\lambda}} \| s_\theta) &= \int s_{\theta,\hat{\lambda}}(y) \log \frac{s_{\theta,\hat{\lambda}}(y)}{s_\theta(y)} \, dy \\
&= \int s_{\theta,\hat{\lambda}}(y) \log \frac{s_\theta(y) \exp(\hat{\lambda}^T h(y) - \psi(\theta,\hat{\lambda}))}{s_\theta(y)} \, dy \\
&= \hat{\lambda}^T \hat{h}(\theta,\hat{\lambda}) - \psi(\theta,\hat{\lambda}) \\
&= \hat{\lambda}^T \bar{h}_N - \psi(\theta,\hat{\lambda}).
\end{aligned} \tag{3.37}$$

Another possibility is to utilize the following identity

$$\begin{aligned}
0 &= \min_\lambda D(s_{\theta,\hat{\lambda}} \| s_{\theta,\lambda}) \\
&= \min_\lambda \int s_{\theta,\hat{\lambda}}(y) \log \frac{s_{\theta,\hat{\lambda}}(y)}{s_\theta(y) \exp(\lambda^T h(y,z) - \psi(\theta,\lambda))} \, dy \\
&= D(s_{\theta,\hat{\lambda}} \| s_\theta) - \max_\lambda \left(\lambda^T \hat{h}(\theta,\hat{\lambda}) - \psi(\theta,\lambda)\right) \\
&= D(s_{\theta,\hat{\lambda}} \| s_\theta) - \max_\lambda \left(\lambda^T \bar{h}_N - \psi(\theta,\lambda)\right).
\end{aligned}$$

As a result, we have

$$\boxed{D(s_{\theta,\hat{\lambda}} \| s_\theta) = \max_\lambda \left(\lambda^T \bar{h}_N - \psi(\theta,\lambda)\right).} \tag{3.38}$$

Hence, the minimum K-L distance is given by maximizing $\lambda^T \bar{h}_N - \psi(\theta,\lambda)$ over $\lambda \in \mathbb{R}^n$. Note that the above expression is known as a Legendre-Fenchel transform (B.5) of $\psi(\theta,\lambda)$ for given θ.

Enveloping Exponential Family. Taking (3.31) and (3.38) together, we have

$$K(r_N{:}s_\theta) = K(r_N{:}s_{\theta,\hat{\lambda}}) + \max_\lambda \left(\lambda^T \bar{h}_N - \psi(\theta,\lambda)\right). \tag{3.39}$$

The practical importance of the Pythagorean relationship (3.31) is that it enables us to evaluate inaccuracy $K(r_N{:}s_\theta)$, with precision up to an additive constant, *without* knowledge of $r_N(y)$ *provided* $K(r_N{:}s_{\theta,\hat{\lambda}})$ is independent of θ for every $r_N(y)$.

The latter means that for every θ, θ' and every $r_N(y)$, it holds

$$K(r_N{:}s_{\theta,\hat{\lambda}}) = K(r_N{:}s_{\theta',\hat{\lambda}'})$$

where $\hat{\lambda}$, $\hat{\lambda}'$ are such that $\hat{h}(\theta,\hat{\lambda}) = \hat{h}(\theta',\hat{\lambda}') = \bar{h}_N$. This condition is to be satisfied for every $r_N(y)$, including

$$r_N(y) = \delta(y-a), \quad a \in \mathcal{Y},$$

which implies

$$s_{\theta,\hat{\lambda}}(y) = s_{\theta',\hat{\lambda}'}(y).$$

Hence, the h-projections of any $r_N(y)$ onto $S_{\theta;h}$ and $S_{\theta';h}$ coincide. But this may happen only if the exponential families $S_{\theta;h}$ and $S_{\theta';h}$ coincide as a whole. If this is the case for every $\theta \in \mathcal{T}$, the model family can be parametrized so that for every θ there exists $\lambda(\theta)$ such that

$$s_\theta(y) = s_{\theta_0}(y) \exp\left(\lambda^T(\theta) h(y) - \psi(\lambda(\theta))\right) \tag{3.40}$$

where $s_{\theta_0}(y)$ is a fixed density from the model family S.

To sum up, the inaccuracy $K(r_N{:}s_{\theta,\hat{\lambda}})$ is independent of θ provided $s_\theta(y)$ for every θ belongs to an exponential family with a fixed origin $s_{\theta_0}(y)$ and a canonical statistic $h(y)$. The choice of $h(y)$ is, however, our choice. If we choose $h(y)$ as a canonical statistic of an arbitrary exponential family that *envelops* the model family S, (3.40) is satisfied by definition and $K(r_N{:}s_{\theta,\hat{\lambda}})$ is independent of θ.

Under the condition (3.40), it follows from (3.39) that

$$\boxed{K(r_N{:}s_\theta) = C + \max_\lambda \left(\lambda^T \bar{h}_N - \psi(\theta,\lambda)\right)} \tag{3.41}$$

where C is a constant independent of θ.

Controlled Dynamic Systems: *View via Conditional Densities*

Assumptions. We consider a family of conditional density functions

$$S \stackrel{\triangle}{=} \left\{s_\theta(y|z) : \theta \in \mathcal{T}\right\}$$

that satisfies the following *regularity conditions*

(a) $\mathcal{T}$ is an open set in $\mathbb{R}^{\dim \theta}$,

(b) $s_\theta(y|z) > 0$ for all $y \in \mathcal{Y}$, $z \in \mathcal{Z}$, $\theta \in \mathcal{T}$,

(c) $s_\theta(y|z)$ is continuously differentiable in y and z for all $\theta \in \mathcal{T}$.

We try to prove under the above assumptions a direct analogy of (3.41) for conditional densities, namely that if $\mathcal{S}$ can be imbedded in an n-dimensional conditional exponential family then it is sufficient to know the minimum conditional inaccuracy projection of the empirical density $r_N(y,z)$ onto the exponential family in order to restore the conditional inaccuracy $\bar{K}(r_N:s_\theta)$ with precision up to an additive constant. We find, however, that the above is not true until the enveloping exponential family is "jointly" exponential in both y and z.

Conditional Exponential Family. Let $s_\theta(y|z)$ be an arbitrary fixed density from $\mathcal{S}$. We construct an exponential family $\bar{\mathcal{S}}_{\theta;f}$ composed of the conditional densities

$$s_{\theta,\tau}(y|z) = s_\theta(y|z) \exp\left(\tau^T f(y;z) - \psi(\theta,\tau;z)\right) \qquad (3.42)$$

where $\tau \in \mathbb{R}^{n'}$ is a natural or canonical parameter of the family, $f:\mathcal{Y} \times \mathcal{Z} \to \mathbb{R}^{n'}$ is a given vector function (canonical statistic) of single observation (y,z) and $\psi(\theta,\tau;z)$ is logarithm of the normalizing divisor

$$\psi(\theta,\tau;z) = \log \int s_\theta(y|z) \exp\left(\tau^T f(y;z)\right) \mathrm{d}y. \qquad (3.43)$$

It is assumed that the functions $f_0(y;z) \equiv 1$, $f_1(y;z)$, $\ldots$, $f_{n'}(y;z)$ are linearly independent modulo a function $C(z)$ of z only, i.e., there is no vector $\tau \neq 0$ such that $\tau^T f(y;z) = C(z)$ where $C(z)$ is an arbitrary function of z. The assumption implies a one-to-one correspondence of τ and $s_{\theta,\tau}(y|z)$. The dimension of the conditional exponential family $\bar{\mathcal{S}}_{\theta;f}$ is then n'.

The parameter τ is assumed to run through all values from $\mathbb{R}^{n'}$ for which the normalizing divisor is finite

$$\exp\left(\psi(\theta,\tau;z)\right) < \infty$$

for every z. It can be shown by Hölder inequality (A.31) that the set of all such values of τ, denoted $\bar{N}_\theta$, is convex and $\psi(\theta,\tau;z)$ is for every $z \in \mathcal{Z}$ a convex function of τ over $\bar{N}_\theta$.

In the sequel, $\bar{\mathcal{S}}_{\theta;f}$ is assumed to be the maximal family of densities that can be expressed as (3.42) for some τ.

Conditional f-Projection. Suppose that a sample y^{N+m}, u^{N+m} is given with the empirical density $r_N(y,z)$. The necessary condition for $\hat{\tau}$ to minimize the conditional inaccuracy (to maximize likelihood) is

$$0 = \nabla_\tau \bar{K}(r_N:s_{\theta,\hat{\tau}})$$

$$= \int\!\!\int r_N(y,z)\left(-f(y;z) + \int s_{\theta,\hat{\tau}}(y|z) f(y;z)\,\mathrm{d}y\right)\mathrm{d}y\,\mathrm{d}z$$

$$= -\int\!\!\int r_N(y,z) f(y;z)\,\mathrm{d}y\,\mathrm{d}z + \int\!\!\int s_{\theta,\hat{\tau}}(y|z)\,\bar{r}_N(z) f(y;z)\,\mathrm{d}y\,\mathrm{d}z.$$

The condition thus reads

$$\iint s_{\theta,\hat{\tau}}(y|z)\,\tilde{r}_N(z)\,f(y;z)\,dy\,dz = \iint r_N(y,z)\,f(y;z)\,dy\,dz \qquad (3.44)$$

where $\tilde{r}_N(z)$ stands for the marginal empirical density

$$\tilde{r}_N(z) = \int r_N(y,z)\,dy.$$

The density $s_{\theta,\hat{\tau}}(y|z)$ that satisfies the condition (3.44) will be called a *conditional f-projection* of $r_N(y,z)$ onto $\check{S}_{\theta;f}$. Introducing the notation

$$\bar{f}_N \triangleq \iint r_N(y,z)\,f(y;z)\,dy\,dz = \frac{1}{N}\sum_{k=m+1}^{N+m} f(y_k;z_k) \qquad (3.45)$$

$$\hat{f}(\theta,\tau;z) \triangleq \int s_{\theta,\tau}(y|z)\,f(y;z)\,dy \qquad (3.46)$$

we can write the condition (3.44) as follows

$$E_N\big(\hat{f}(\theta,\hat{\tau};Z)\big) = \bar{f}_N.$$

We denote the set of all densities $r(y,z)$ with the same f-projection as

$$\mathcal{R}_N \triangleq \Big\{ r(y,z) : \iint r(y,z)\,f(y;z)\,dy\,dz = \bar{f}_N,\ \iint r(y,z)\,dy\,dz = 1,\ r(y,z) \geq 0 \Big\}.$$

Note that the value $\hat{f}(\theta,\tau;z)$ coincides with the gradient of logarithm of the normalizing divisor $\psi(\theta,\tau;z)$ with respect to τ

$$\begin{aligned}
\nabla_\tau \psi(\theta,\tau;z) &= \nabla_\tau \log \int s_\theta(y|z)\,\exp\big(\tau^T f(y;z)\big)\,dy \\
&= \frac{\int s_\theta(y|z)\,\nabla_\tau \exp\big(\tau^T f(y;z)\big)\,dy}{\int s_\theta(y|z)\,\exp\big(\tau^T f(y;z)\big)\,dy} \\
&= \int \frac{s_\theta(y|z)\,\exp\big(\tau^T f(y;z)\big)}{\int s_\theta(y|z)\,\exp\big(\tau^T f(y;z)\big)\,dy}\,f(y;z)\,dy \\
&= \int s_{\theta,\tau}(y|z)\,f(y;z)\,dy \\
&= \hat{f}(\theta,\tau;z) \qquad (3.47)
\end{aligned}$$

Conditional Pythagorean Relationship. Let $s_{\theta,\tau}(y|z)$ be exponential (3.42) and $\hat{\tau}$ satisfy (3.44). Then we can write

$$\bar{K}(r_N:s_\theta) - \bar{K}(r_N:s_{\theta,\hat{\tau}})$$

$$= \iint r_N(y,z) \log \frac{s_{\theta,\hat{\tau}}(y|z)}{s_\theta(y|z)}\,dy\,dz$$

$$= \iint r_N(y,z) \log \frac{s_\theta(y|z) \exp(\hat{\tau}^T f(y;z) - \psi(\theta,\hat{\tau};z))}{s_\theta(y|z)} \, dy \, dz$$

$$= \hat{\tau}^T \left(\iint r_N(y,z) f(y;z) \, dy \, dz \right) - \psi(\theta,\hat{\tau};z)$$

$$= \hat{\tau}^T \left(\iint s_{\theta,\hat{\tau}}(y|z) \tilde{r}_N(z) f(y;z) \, dy \, dz \right) - \psi(\theta,\hat{\tau};z)$$

$$= \iint s_{\theta,\hat{\tau}}(y|z) \tilde{r}_N(z) \log \frac{s_\theta(y|z) \exp(\hat{\tau}^T f(y;z) - \psi(\theta,\hat{\tau}))}{s_\theta(y|z)} \, dy \, dz$$

$$= \iint s_{\theta,\hat{\tau}}(y|z) \tilde{r}_N(z) \log \frac{s_{\theta,\hat{\tau}}(y|z)}{s_\theta(y|z)} \, dy \, dz$$

$$= \bar{D}(s_{\theta,\hat{\tau}} \| s_\theta | \tilde{r}_N).$$

Hence, we obtain a conditional version of the Pythagorean relationship (cf. Fig. 3.10)

$$\bar{K}(r_N : s_\theta) = \bar{K}(r_N : s_{\theta,\hat{\tau}}) + \bar{D}(s_{\theta,\hat{\tau}} \| s_\theta | \tilde{r}_N). \tag{3.48}$$

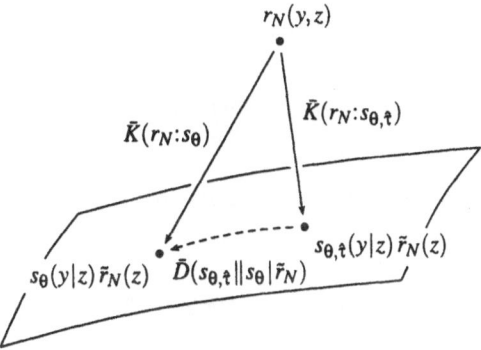

Fig. 3.10. Pythagorean-like decomposition of conditional inaccuracy for dependent observations.

Minimum Conditional Inaccuracy Projection. Suppose that the conditional inaccuracy $\bar{K}(r_N : s_{\theta,\hat{\tau}})$ relative to the f-projection $s_{\theta,\hat{\tau}}(y|z)$ is finite

$$\bar{K}(r_N : s_{\theta,\hat{\tau}}) < \infty, \tag{3.49}$$

Then the following Pythagorean relationship holds for every $\tau \in \tilde{N}_\theta$

$$\bar{K}(r_N : s_{\theta,\tau}) - \bar{K}(r_N : s_{\theta,\hat{\tau}})$$

$$= \iint r_N(y,z) \log \frac{s_{\theta,\hat{\tau}}(y|z)}{s_{\theta,\tau}(y|z)} \, dy \, dz$$

$$= \iint r_N(y,z) \log \frac{s_\theta(y|z) \exp(\hat{t}^T f(y;z) - \psi(\theta,\hat{t};z))}{s_\theta(y|z) \exp(\tau^T f(y;z) - \psi(\theta,\tau;z))} \, dy \, dz$$

$$= (\hat{t} - \tau)^T \left(\iint r_N(y,z) f(y;z) \, dy \, dz \right) - \psi(\theta,\hat{t};z) + \psi(\theta,\tau;z)$$

$$= (\hat{t} - \tau)^T \left(\iint s_{\theta,\hat{t}}(y|z) \tilde{r}_N(z) f(y;z) \, dy \, dz \right) - \psi(\theta,\hat{t};z) + \psi(\theta,\tau;z)$$

$$= \iint s_{\theta,\hat{t}}(y|z) \tilde{r}_N(z) \log \frac{s_\theta(y|z) \exp(\hat{t}^T f(y;z) - \psi(\theta,\hat{t};z))}{s_\theta(y|z) \exp(\tau^T f(y;z) - \psi(\theta,\tau;z))} \, dy \, dz$$

$$= \iint s_{\theta,\hat{t}}(y|z) \tilde{r}_N(z) \log \frac{s_{\theta,\hat{t}}(y|z)}{s_{\theta,\tau}(y|z)} \, dy \, dz$$

$$= \bar{D}(s_{\theta,\hat{t}} \| s_{\theta,\tau} | \tilde{r}_N).$$

Since the conditional K-L distance $\bar{D}(s_{\theta,\hat{t}} \| s_{\theta,\tau} | \tilde{r}_N)$ is nonnegative by (3.17) and the conditional inaccuracy $\bar{K}(r_N : s_{\theta,\hat{t}})$ was assumed finite in (3.49), we have

$$\bar{K}(r_N : s_{\theta,\tau}) \geq \bar{K}(r_N : s_{\theta,\hat{t}})$$

with equality if and only if $s_{\theta,\tau}(y|z) \tilde{r}_N(z) = s_{\theta,\hat{t}}(y) \tilde{r}_N(z)$ almost everywhere. Thus, the f-projection $s_{\theta,\hat{t}}(y)$ is a solution to the minimum conditional inaccuracy problem

$$\boxed{\bar{K}(r_N : s_{\theta,\hat{t}}) = \min_{\tau \in \tilde{N}_\theta} \bar{K}(r_N : s_{\theta,\tau}).} \tag{3.50}$$

Note that the uniqueness of $s_{\theta,\hat{t}}(y|z) \tilde{r}_N(z)$ does not imply necessarily the uniqueness of $s_{\theta,\hat{t}}(y|z)$ itself.

Minimum Conditional K-L Distance Projection. There is no obvious analogue to the dual interpretation of f-projection shown in the case of independent observations. Assume that the conditional K-L distance of the f-projection $s_{\theta,\hat{t}}(y|z)$ from the model density $s_\theta(y|z)$ is finite

$$\bar{D}(s_{\theta,\hat{t}} \| s_\theta | \tilde{r}_N) < \infty. \tag{3.51}$$

Then there are $r(y,z) \in \mathcal{R}_N$ such that $\bar{D}(r \| s_\theta) < \infty$. Given such $r(y,z) \in \mathcal{R}_N$, we fail generally to prove a Pythagorean relationship when $r(y,z) \neq r_N(y,z)$

$$\bar{D}(r \| s_\theta) - \bar{D}(r \| s_{\theta,\hat{t}})$$

$$= \iint r(y,z) \log \frac{s_{\theta,\hat{t}}(y|z)}{s_\theta(y|z)} \, dy \, dz$$

$$= \iint r(y,z) \log \frac{s_\theta(y|z) \exp(\hat{t}^T f(y;z) - \psi(\theta,\hat{t};z))}{s_\theta(y|z)} \, dy \, dz$$

$$= \hat{t}^T \left(\iint r(y,z) f(y;z) \, dy \, dz \right) - \psi(\theta,\hat{t};z)$$

$$\neq \hat{t}^T \left(\iint s_{\theta,\hat{t}}(y|z) \tilde{r}(z) f(y;z) \, dy \, dz \right) - \psi(\theta,\hat{t};z)$$

$$= \iint s_{\theta,\hat{t}}(y|z)\,\tilde{r}(z) \log \frac{s_{\theta}(y|z)\exp(\hat{t}^T f(y;z) - \psi(\theta,\hat{t};))}{s_{\theta}(y|z)}\,dy\,dz$$

$$= \iint s_{\theta,\hat{t}}(y|z)\,\tilde{r}(z) \log \frac{s_{\theta,\hat{t}}(y|z)}{s_{\theta}(y|z)}\,dy\,dz$$

$$= \bar{D}(s_{\theta,\hat{t}} \| s_{\theta} | \tilde{r}). \tag{3.52}$$

The lack of analogy with the case of independent observations follows from the fact that there may be $r(y,z) \in \mathcal{R}_N$ such that

$$\iint s_{\theta,\hat{t}}(y|z)\,\tilde{r}(z)\,f(y;z)\,dy\,dz \neq \iint s_{\theta,\hat{t}}(y|z)\,\tilde{r}_N(z)\,f(y;z)\,dy\,dz = \tilde{f}_N.$$

The equality appears above only when the set $\mathcal{R}_N$ is further restricted, i.e., the statistic $E_N(f(Y;Z))$ is properly complemented as shown below.

Minimum Conditional K-L Distance. One possibility of evaluating the conditional K-L distance $\bar{D}(s_{\theta,\hat{t}} \| s_{\theta} | \tilde{r}_N)$ is to determine the f-projection $s_{\theta,\hat{t}}(y|z)$ explicitly and substitute it in $\bar{D}(s_{\theta,\hat{t}} \| s_{\theta} | \tilde{r}_N)$. This results in

$$\bar{D}(s_{\theta,\hat{t}} \| s_{\theta} | \tilde{r}_N) = \iint s_{\theta,\hat{t}}(y|z)\,\tilde{r}_N(z) \log \frac{s_{\theta,\hat{t}}(y|z)}{s_{\theta}(y|z)}\,dy\,dz$$

$$= \iint s_{\theta,\hat{t}}(y|z)\,\tilde{r}_N(z) \log \frac{s_{\theta}(y|z)\exp(\hat{t}^T f(y;z) - \psi(\theta,\hat{t};z))}{s_{\theta}(y|z)}\,dy\,dz$$

$$= \hat{t}^T E_N(\hat{f}(\theta,\hat{t};Z)) - E_N(\psi(\theta,\hat{t};Z))$$

$$= \hat{t}^T \tilde{f}_N - E_N(\psi(\theta,\hat{t};Z)) \tag{3.53}$$

Another way of computation is to make use of the identity

$$0 = \min_{\tau} \bar{D}(s_{\theta,\hat{t}} \| s_{\theta,\tau} | \tilde{r}_N)$$

$$= \min_{\tau} \left(\iint s_{\theta,\hat{t}}(y|z)\,\tilde{r}_N(z) \log \frac{s_{\theta,\hat{t}}(y|z)}{s_{\theta}(y|z)\exp(\tau^T f(y;z) - \psi(\theta,\tau;z))}\,dy\,dz \right)$$

$$= \bar{D}(s_{\theta,\hat{t}} \| s_{\theta} | \tilde{r}_N) - \max_{\tau} \left(\tau^T E_N(\hat{f}(\theta,\hat{t};Z)) - E_N(\psi(\theta,\tau;Z)) \right)$$

$$= \bar{D}(s_{\theta,\hat{t}} \| s_{\theta} | \tilde{r}_N) - \max_{\tau} \left(\tau^T \tilde{f}_N - E_N(\psi(\theta,\tau;Z)) \right).$$

As a result, we have

$$\boxed{\bar{D}(s_{\theta,\hat{t}} \| s_{\theta} | \tilde{r}_N) = \max_{\tau} \left(\tau^T \tilde{f}_N - E_N(\psi(\theta,\tau;Z)) \right).} \tag{3.54}$$

Thus, the minimum conditional K-L distance can be determined by maximizing $\tau^T \tilde{f}_N - E_N(\psi(\theta,\tau;Z))$ over $\tau \in \mathbb{R}^{n'}$.

Enveloping Conditional Exponential Family. Taking (3.48) and (3.54) together, we have

$$\bar{K}(r_N{:}s_\theta) = \bar{K}(r_N{:}s_{\theta,\hat{t}}) + \max_\tau \left(\tau^T \bar{f}_N - E_N(\psi(\theta,\tau;Z)) \right). \tag{3.55}$$

Owing to the Pythagorean relationship, it is possible to evaluate the conditional inaccuracy $\bar{K}(r_N{:}s_\theta)$, with precision up to an additive constant, *without complete* knowledge of $r_N(y,z)$ provided $\bar{K}(r_N{:}s_{\theta,\hat{t}})$ is independent of θ for every $r_N(y,z)$.

The latter means that for every θ, θ' and every $r_N(y,z)$, it holds

$$\bar{K}(r_N{:}s_{\theta,\hat{t}}) = \bar{K}(r_N{:}s_{\theta',\hat{t}'})$$

where $\hat{t}$, $\hat{t}'$ are such that $E_N(\hat{f}(\theta,\hat{t};Z)) = E_N(\hat{f}(\theta',\hat{t}';Z)) = \bar{f}_N$. This condition is to be satisfied for every $r_N(y,z)$, including

$$r_N(y,z) = \delta(y-a, z-b), \quad a \in \mathcal{Y}, \quad b \in \mathcal{Z}$$

which implies

$$s_{\theta,\hat{t}}(y|z) = s_{\theta',\hat{t}'}(y|z).$$

Hence, the f-projections of any $r_N(y,z)$ onto $\bar{\mathcal{S}}_{\theta;f}$ and $\bar{\mathcal{S}}_{\theta';f}$ coincide. But this may happen only if the exponential families $\bar{\mathcal{S}}_{\theta;f}$ and $\bar{\mathcal{S}}_{\theta';f}$ coincide as a whole. If this is the case for all $\theta \in \mathcal{T}$, the model family can be parametrized so that for every θ there exists $\tau(\theta)$ such that

$$s_\theta(y|z) = s_{\theta_0}(y|z) \exp\left(\tau^T(\theta) f(y;z) - \psi(\tau(\theta);z) \right) \tag{3.56}$$

where $s_{\theta_0}(y|z)$ is a fixed density from the model family $\mathcal{S}$.

We can sum up that the inaccuracy $\bar{K}(r_N{:}s_{\theta,\hat{t}})$ is independent of θ provided $s_\theta(y|z)$ for every θ belongs to a conditional exponential family with a fixed origin $s_{\theta_0}(y|z)$ and a canonical statistic $f(y;z)$. Thus, if we choose $f(y;z)$ as a canonical statistic of any conditional exponential family *enveloping* the model family $\mathcal{S}$, (3.56) is satisfied by definition and $\bar{K}(r_N{:}s_{\theta,\hat{t}})$ is independent of θ.

Under the condition (3.56), we have from (3.55)

$$\boxed{\bar{K}(r_N{:}s_\theta) = C + \max_\tau \left(\tau^T \bar{f}_N - E_N(\psi(\theta,\tau;Z)) \right)} \tag{3.57}$$

where C is constant independent of θ.

Marginal Inaccuracy. It is easy to see that even under the assumption (3.56), we do not get rid of the dependence on $r_N(y,z)$ completely. Implementation of (3.54) requires computation of the marginal empirical expectation $E_N(\psi(\theta,\tau;Z))$. Since the parameters θ and τ are unknown, the empirical expectation would have to be performed for every possible value of them. This is, of course, impracticable except a special case discussed below when the function $\psi(\theta,\tau;z)$ can be factorized so to separate (θ,τ) and z.

To solve the general case, we exploit the fact that evaluation of $E_N\big(-\psi(\theta,\tau;Z)\big)$ is largely analogous to evaluation of $E_N\big(-\log s_\theta(Y|Z)\big)$. Introducing a positive but unnormalized function

$$w_{\theta,\tau}(z) = \exp\big(\psi(\theta,\tau;z)\big) \tag{3.58}$$

and extending the definition of inaccuracy to

$$K(\tilde{r}_N{:}w_{\theta,\tau}) = \int \tilde{r}_N(z) \log \frac{1}{w_{\theta,\tau}(z)}\, dz,$$

we face the problem to compute

$$E_N\big(-\psi(\theta,\tau;Z)\big) = K(\tilde{r}_N{:}w_{\theta,\tau}) \tag{3.59}$$

which is formally identical to the problem we solved for independent observations.

Let us stress again that $w_{\theta,\tau}(z)$ is *not* a probability density function because of the lack of normalization. Yet, we can easily deduce an analogue of the Pythagorean relationship that holds even in the unnormalized case.

Marginal Exponential Family. We introduce an exponential family $\mathcal{W}_{\theta,\tau;g}$ composed of density functions

$$w_{\theta,\tau,\mu}(z) = w_{\theta,\tau}(z) \exp\big(\mu^T g(z) - \psi(\theta,\tau,\mu)\big) \tag{3.60}$$

where $\mu \in \mathbb{R}^{n''}$ is a natural or canonical parameter of the family, $g\colon \mathcal{Z} \to \mathbb{R}^{n''}$ is a given function (canonical statistic) of single observation z and $\psi(\theta,\tau,\mu)$ is logarithm of the normalizing divisor which owing to (3.58) and (3.43) takes the form

$$\begin{aligned}
\psi(\theta,\tau,\mu) &= \log \int w_{\theta,\tau}(z) \exp\big(\mu^T g(z)\big) dz \\
&= \log \int \exp\big(\psi(\theta,\tau;z) + \mu^T g(z)\big) dz \\
&= \log \iint s_\theta(y|z) \exp\big(\tau^T f(y;z) + \mu^T g(z)\big) dy\, dz.
\end{aligned}$$

It is assumed that the functions $g_0(z) \equiv 1, g_1(z), \ldots, g_{n''}(z)$ are linearly independent which implies a one-to-one correspondence of μ and $w_{\theta,\tau,\mu}(z)$. The dimension of the marginal exponential family is then n''.

The parameter μ is assumed to run through all values from $\mathbb{R}^{n''}$ for which the normalizing divisor is finite

$$\exp\big(\psi(\theta,\tau,\mu)\big) < \infty.$$

It can shown by Hölder inequality (A.31) that the set of all such values of μ, denoted $\mathcal{N}_{\theta,\tau}$, is convex and $\psi(\theta,\tau,\mu)$ is a convex function of μ on $\mathcal{N}_{\theta,\tau}$.

Hence, $\mathcal{W}_{\theta,\tau;g}$ is assumed to be the maximal family of densities that can be expressed as (3.60) for some μ.

We follow now essentially the same sequence of steps as we did in the case of independent observations.

Marginal g-Projection. Suppose a sample y^{N+m}, u^{N+m} is given with the marginal empirical density $\tilde{r}_N(z)$. The necessary condition for $\hat{\mu}$ to minimize the marginal (unnormalized) inaccuracy is

$$0 = \nabla_\mu K(\tilde{r}_N : w_{\theta,\tau,\hat{\mu}})$$

$$= \int \tilde{r}_N(z) \left(-g(z) + \int w_{\theta,\tau,\hat{\mu}}(z) g(z) \, dz \right)$$

$$= -\int \tilde{r}_N(z) g(z) \, dz + \int w_{\theta,\tau,\hat{\mu}}(z) g(z) \, dz,$$

that is

$$\boxed{\int w_{\theta,\tau,\hat{\mu}}(z) g(z) \, dz = \int \tilde{r}_N(z) g(z) \, dz.} \tag{3.61}$$

The density $w_{\theta,\tau,\hat{\mu}}(z)$ that satisfies the condition (3.61) will be called a *marginal g-projection* of $\tilde{r}_N(z)$ onto $\mathcal{W}_{\theta,\tau;g}$. Introducing the notation

$$\bar{g}_N \triangleq \int \tilde{r}_N(z) g(z) \, dz = \frac{1}{N} \sum_{k=m+1}^{N+m} g(z_k), \tag{3.62}$$

$$\hat{g}(\theta,\tau,\mu) \triangleq \int w_{\theta,\tau,\mu}(z) g(z) \, dz, \tag{3.63}$$

we can write (3.61) as

$$\hat{g}(\theta,\tau,\hat{\mu}) = \bar{g}_N.$$

Note again that $\hat{g}(\theta,\tau,\mu)$ coincides with the gradient of logarithm of the normalizing divisor $\psi(\theta,\tau,\mu)$ with respect to μ

$$\nabla_\mu \psi(\theta,\tau,\mu) = \nabla_\mu \log \int w_{\theta,\tau}(z) \exp\left(\mu^T g(z)\right) dz$$

$$= \frac{\int w_{\theta,\tau}(z) \nabla_\mu \exp\left(\mu^T g(z)\right) dz}{\int w_{\theta,\tau}(z) \exp\left(\mu^T g(z)\right) dz}$$

$$= \int \frac{w_{\theta,\tau}(z) \exp\left(\mu^T g(z)\right)}{\int w_{\theta,\tau}(z) \exp\left(\mu^T g(z)\right)} g(z) \, dz$$

$$= \int w_{\theta,\tau,\mu}(z) g(z) \, dz$$

$$= \hat{g}(\theta,\tau,\mu). \tag{3.64}$$

Marginal Pythagorean Relationship. Let $w_{\theta,\tau,\mu}(z)$ be exponential (3.60) and $\hat{\mu}$ satisfy (3.61). Then

$$K(\tilde{r}_N : w_{\theta,\tau}) - K(\tilde{r}_N : w_{\theta,\tau,\hat{\mu}})$$

$$= \int \tilde{r}_N(z) \log \frac{w_{\theta,\tau,\hat{\mu}}(z)}{w_{\theta,\tau}(z)} \, dz$$

$$= \int \tilde{r}_N(z) \log \frac{w_{\theta,\tau}(z) \exp\left(\hat{\mu}^T g(z) - \psi(\theta,\tau,\hat{\mu})\right)}{w_{\theta,\tau}(z)} \, dz$$

$$= \hat{\mu}^T \left(\int \bar{r}_N(z) g(z) \, dz \right) - \psi(\theta, \tau, \hat{\mu})$$

$$= \hat{\mu}^T \left(\int w_{\theta,\tau,\hat{\mu}}(y) g(z) \, dz \right) - \psi(\theta, \tau, \hat{\mu})$$

$$= \int w_{\theta,\tau,\hat{\mu}}(z) \log \frac{w_{\theta,\tau}(z) \exp\left(\hat{\mu}^T g(z) - \psi(\theta, \tau, \hat{\mu}) \right)}{w_{\theta,\tau}(z)} \, dz$$

$$= \int w_{\theta,\tau,\hat{\mu}}(z) \log \frac{w_{\theta,\tau,\hat{\mu}}(z)}{w_{\theta,\tau}(z)} \, dz$$

$$= D(w_{\theta,\tau,\hat{\mu}} \| w_{\theta,\tau}).$$

As a result, we obtain a marginal (unnormalized) version of the Pythagorean relationship

$$K(\bar{r}_N : w_{\theta,\tau}) = K(\bar{r}_N : w_{\theta,\tau,\hat{\mu}}) + D(w_{\theta,\tau,\hat{\mu}} \| w_{\theta,\tau}). \tag{3.65}$$

Minimum Marginal K-L Distance. There are two possible ways again of evaluating the marginal (unnormalized) K-L distance $D(w_{\theta,\tau,\hat{\mu}} \| w_{\theta,\tau})$. One possibility is to determine the g-projection $w_{\theta,\tau,\hat{\mu}}$ explicitly and substitute it in $D(w_{\theta,\tau,\hat{\mu}} \| w_{\theta,\tau})$. This yields

$$D(w_{\theta,\tau,\hat{\mu}} \| w_{\theta,\tau}) = \int w_{\theta,\tau,\hat{\mu}}(z) \log \frac{w_{\theta,\tau,\hat{\mu}}(z)}{w_{\theta,\tau}(z)} \, dz$$

$$= \int w_{\theta,\tau,\hat{\mu}}(z) \log \frac{w_{\theta,\tau}(z) \exp\left(\hat{\mu}^T g(z) - \psi(\theta, \tau, \hat{\mu}) \right)}{w_{\theta,\tau}(z)} \, dz$$

$$= \hat{\mu}^T \hat{g}(\theta, \tau, \hat{\mu}) - \psi(\theta, \tau, \hat{\mu})$$

$$= \hat{\mu}^T \bar{g}_N - \psi(\theta, \tau, \hat{\mu}). \tag{3.66}$$

An alternative way of calculation follows from the identity

$$0 = \min_{\mu} D(w_{\theta,\tau,\hat{\mu}} \| w_{\theta,\tau,\mu})$$

$$= \min_{\mu} \int w_{\theta,\tau,\hat{\mu}}(z) \log \frac{w_{\theta,\tau,\hat{\mu}}(z)}{w_{\theta,\tau}(z) \exp\left(\mu^T g(z) - \psi(\theta, \tau, \mu) \right)} \, dz$$

$$= D(w_{\theta,\tau,\hat{\mu}} \| w_{\theta,\tau}) - \max_{\tau} \left(\mu^T \hat{g}(\theta, \tau, \hat{\mu}) - \psi(\theta, \tau, \mu) \right)$$

$$= D(w_{\theta,\tau,\hat{\mu}} \| w_{\theta,\tau}) - \max_{\tau} \left(\mu^T \bar{g}_N - \psi(\theta, \tau, \mu) \right).$$

As a result, we have

$$D(w_{\theta,\tau,\hat{\mu}} \| w_{\theta,\tau}) = \max_{\mu} \left(\mu^T \bar{g}_N - \psi(\theta, \tau, \mu) \right). \tag{3.67}$$

The marginal (unnormalized) K-L distance can thus be found by maximizing $\mu^T \bar{g}_N - \psi(\theta, \tau, \mu)$ over $\mu \in \mathbb{R}^{n''}$.

Enveloping Marginal Exponential Family. Taking (3.65) and (3.67) together, we obtain

$$K(\tilde{r}_N : w_{\theta,\tau}) = K(\tilde{r}_N : w_{\theta,\tau,\hat{\mu}}) + \max_{\mu} \left(\mu^T \bar{g}_N - \psi(\theta,\tau,\mu) \right) \qquad (3.68)$$

Thanks to the (unnormalized) Pythagorean relationship (3.65), the marginal (unnormalized) inaccuracy $K(\tilde{r}_N : w_{\theta,\tau})$ can be evaluated, with precision up to an additive constant, *without* knowledge of $\tilde{r}_N(z)$ *provided* $K(\tilde{r}_N : w_{\theta,\tau,\hat{\mu}})$ is independent of (θ,τ) for every $\tilde{r}_N(z)$.

The latter means that for every (θ,τ), (θ',τ') and every $\tilde{r}_N(z)$, it holds

$$K(\tilde{r}_N : w_{\theta,\tau,\hat{\mu}}) = K(\tilde{r}_N : w_{\theta',\tau',\hat{\mu}'})$$

where $\hat{\mu}$, $\hat{\mu}'$ are such that $\hat{g}(\theta,\tau,\hat{\mu}) = \hat{g}(\theta',\tau',\hat{\mu}') = \bar{g}_N$. This condition is to be satisfied for every $\tilde{r}_N(z)$, including

$$\tilde{r}_N(z) = \delta(z-b), \quad b \in \mathcal{Z},$$

which implies

$$w_{\theta,\tau,\hat{\mu}}(y) = w_{\theta',\tau',\hat{\mu}'}(y).$$

Hence, the g-projections of any $\tilde{r}_N(z)$ onto $\mathcal{W}_{\theta,\tau;g}$ and $\mathcal{W}_{\theta',\tau';g}$ coincide. But this may happen only if the exponential families $\mathcal{W}_{\theta,\tau;g}$ and $\mathcal{W}_{\theta',\tau';g}$ coincide as a whole. If this is the case for all θ and τ, the family $\mathcal{W}_{\theta,\tau;g}$ can be parametrized so that for every (θ,τ) there exists $\mu(\theta,\tau)$ such that

$$w_{\theta,\tau}(z) = C w_{\theta_0,\tau_0}(z) \exp\left(\mu^T(\theta,\tau) g(z)\right) \qquad (3.69)$$

where $w_{\theta_0,\tau_0}(z)$ is a fixed function of the form (3.58) and C is a constant independent of z.

Thus, the unnormalized inaccuracy $K(\tilde{r}_N : w_{\theta,\tau,\hat{\mu}})$ is independent of (θ,τ) provided $w_{\theta,\tau}(z)$ for every (θ,τ) belongs to an unnormalized exponential family

$$\left\{ C w_{\theta_0,\tau_0}(z) \exp\left(\mu^T g(z)\right) : C > 0, \; \mu \in \mathbb{R}^{n''} \right\}$$

with a fixed origin $w_{\theta_0,\tau_0}(z)$ and the canonical statistic $g(z)$. Thus, if we choose $g(z)$ as a canonical statistic of any unnormalized exponential family *enveloping* the family of positive functions

$$\{ w_{\theta,\tau}(z) : \theta \in \mathcal{T}, \tau \in \bar{N}_\theta \subset \mathbb{R}^{n'} \},$$

then (3.69) is clearly satisfied and $K(\tilde{r}_N : w_{\theta,\tau,\hat{\mu}})$ is independent of (θ,τ).

Under the condition (3.69), we can write (3.68) as

$$\boxed{K(\tilde{r}_N : w_{\theta,\tau}) = C + \max_{\mu} \left(\mu^T \bar{g}_N - \psi(\theta,\tau,\mu) \right)} \qquad (3.70)$$

where C is a constant independent of θ.

Enveloping Conditional Exponential Family II. Putting (3.57), (3.59) and (3.70) together, we can conclude that under the conditions (3.56) and (3.69), the conditional inaccuracy can be evaluated as

$$\boxed{\bar{K}(r_N{:}s_\theta) = C + \max_{\tau,\mu}\left(\tau^T \bar{f}_N + \mu^T \bar{g}_N - \psi(\theta,\tau,\mu)\right)} \tag{3.71}$$

where C is constant independent of θ.

Note that the conditions (3.56) and (3.69) can equivalently be written so that the following holds for every θ and τ

$$\log \frac{s_\theta(y|z)}{s_{\theta_0}(y|z)} \in \operatorname{span}\left\{1, f_1(y;z), \ldots, f_{n'}(y;z)\right\} + C(z), \tag{3.72}$$

$$\psi(\theta,\tau;z) - \psi(\theta_0,\tau_0;z) \in \operatorname{span}\left\{1, g_1(z), \ldots, g_{n''}(z)\right\}. \tag{3.73}$$

Here $\operatorname{span}\{\cdot\}$ denotes the linear space spanned by given functions and $C(z)$ is an arbitrary function of z. The conditions (3.72) and (3.73) mean that

(a) the model family is imbedded in a conditional exponential family with the canonical statistic $f(y;z)$,

(b) logarithm of the normalizing divisor $\psi(\theta,\tau;z)$ of the enveloping conditional exponential family is separable in (θ,τ) and z,

respectively. Only if both the conditions are satisfied, the conditional inaccuracy $\bar{K}(r_N{:}s_\theta)$ can be calculated *without* complete knowledge of the empirical density $r_N(y,z)$; all we need to store then are the empirical expectations $\bar{f}_N$ and $\bar{g}_N$.

Let us stress that the conditions are very strict; the condition (3.73) is rarely satisfied with finite or sufficiently small n''.

Controlled Dynamic Systems:
View via Joint Densities

Suppose that the condition (3.69) or, equivalently, (3.73) is satisfied, i.e., $K(\bar{r}_N{:}w_{\theta,\tau,\hat{\mu}})$ is independent of (θ,τ) for every $\bar{r}_N(z)$. Then by (3.55) and (3.68) we have

$$\bar{K}(r_N{:}s_\theta) = \bar{K}(r_N{:}s_{\theta,\hat{t}}) + K(\bar{r}_N{:}w_{\theta,\hat{t},\hat{\mu}}) + \max_{\tau,\mu}\left(\tau^T \bar{f}_N + \mu^T \bar{g}_N - \psi(\theta,\tau,\mu)\right). \tag{3.74}$$

The formula (3.74) can be further simplified. First, it holds

$$\bar{K}(r_N{:}s_{\theta,\hat{t}}) + K(\bar{r}_N{:}w_{\theta,\hat{t},\hat{\mu}})$$

$$= \iint r_N(y,z) \log \frac{1}{s_{\theta,\hat{t}}(y|z)}\, dy\, dz + \int \bar{r}_N(z) \log \frac{1}{w_{\theta,\hat{t},\hat{\mu}}(z)}\, dz$$

$$= \iint r_N(y,z) \log \frac{1}{s_{\theta,\hat{t}}(y|z)\, w_{\theta,\hat{t},\hat{\mu}}(z)}\, dy\, dz$$

$$= K(r_N{:}s_{\theta,\hat{t}}\, w_{\theta,\hat{t},\hat{\mu}})$$

where

$$
\begin{aligned}
s_{\theta,\hat{t}}(y|z)\, w_{\theta,\hat{t},\hat{\mu}}(z) &= s_\theta(y|z)\exp\!\big(\hat{t}^T f(y;z) - \psi(\theta,\hat{t};z)\big) \\
&\quad \cdot \exp\!\big(\psi(\theta,\hat{t};z) + \hat{\mu}^T g(z) - \psi(\theta,\hat{t},\hat{\mu})\big) \\
&= s_\theta(y|z)\exp\!\big(\hat{t}^T f(y;z) + \hat{\mu}^T g(z) - \psi(\theta,\hat{t},\hat{\mu})\big).
\end{aligned}
$$

Second, the conditional inaccuracy $\bar{K}(r_N{:}s_\theta)$ can be regarded as the *unnormalized joint inaccuracy* of $r_N(y,z)$ relative to the function $s_\theta(y|z)$

$$
\begin{aligned}
\bar{K}(r_N{:}s_\theta) &= \iint r_N(y,z)\log\frac{1}{s_\theta(y|z)}\,dy\,dz \\
&= K(r_N{:}s_\theta).
\end{aligned}
$$

With this notation, the formula (3.74) can be rewritten as

$$
K(r_N{:}s_\theta) = K(r_N{:}s_{\theta,\hat{t}}\,w_{\theta,\hat{t},\hat{\mu}}) + \max_{\tau,\mu}\big(\tau^T \bar{f}_N + \mu^T \bar{g}_N - \psi(\theta,\tau,\mu)\big).
$$

The formal similarity of the resulting formula with the independent observations case (3.39) is striking and suggests an alternative approach to estimation via *joint* densities of Y and Z.

Joint Exponential Family. Consider a fixed point $s_\theta(y|z)$ for a particular value of θ and construct a *joint* exponential family $S_{\theta;h}$ composed of the joint densities

$$
s_{\theta,\lambda}(y,z) = s_\theta(y|z)\exp\!\big(\lambda^T h(y,z) - \psi(\theta,\lambda)\big) \tag{3.75}
$$

where $\lambda \in \mathbb{R}^n$ is a natural or canonical parameter of the family, $h{:}\,\mathcal{Y}\times\mathcal{Z}\mapsto\mathbb{R}^n$ is a given function (canonical statistic) of (y,z) and

$$
\psi(\theta,\lambda) = \log\iint s_\theta(y|z)\exp\!\big(\lambda^T h(y,z)\big)\,dy\,dz \tag{3.76}
$$

is logarithm of the normalizing divisor.

It is assumed that the functions $h_0(y,z) \equiv 1,\ h_1(y,z),\ \ldots,\ h_n(y,z)$ are linearly independent. Since two densities $s_{\theta,\lambda}(y,z)$ and $s_{\theta,\lambda'}(y,z)$ are equal if and only if the right-hand side of

$$
\log\frac{s_{\theta,\lambda}(y,z)}{s_{\theta,\lambda'}(y,z)} = (\lambda - \lambda')^T h(y,z) - \psi(\theta,\lambda) + \psi(\theta,\lambda')
$$

vanishes, the assumption implies a one-to-one correspondence between the vector parameter λ and the joint density $s_{\theta,\lambda}(y,z)$. The dimension of $S_{\theta;h}$ then equals n.

Normalizing Divisor. The parameter λ is assumed to run through all values from $\mathbb{R}^n$ for which the normalizing divisor is finite

$$
\exp\!\big(\psi(\theta,\lambda)\big) < \infty.
$$

The set of all such values of λ will be denoted by $\mathcal{N}_\theta$. We show that the set $\mathcal{N}_\theta$ is convex. Let λ and λ' be two parameter points for which the normalizing divisor is finite. Then by Hölder inequality (A.31) for every $0 < a < 1$

$$\iint s_\theta(y|z) \exp\left((a\lambda + (1-a)\lambda')^T h(y,z)\right) dy\,dz$$

$$\le \left(\iint s_\theta(y|z) \exp(\lambda^T h(y,z))\,dy\,dz\right)^a$$

$$\cdot \left(\iint s_\theta(y|z) \exp(\lambda'^T h(y,z))\,dy\,dz\right)^{1-a}$$

$$< \infty.$$

By taking logarithms of both sides, we obtain

$$\log \iint s_\theta(y|z) \exp\left((a\lambda + (1-a)\lambda')^T h(y,z)\right) dy\,dz$$

$$\le a \log\left(\iint s_\theta(y|z) \exp(\lambda^T h(y,z))\,dy\,dz\right)$$

$$+ (1-a) \log\left(\iint s_\theta(y|z) \exp(\lambda'^T h(y,z))\,dy\,dz\right).$$

Hence, the function $\psi(\theta, \lambda)$ is a convex function of λ on the convex set $\mathcal{N}_\theta$.

In the sequel, $S_{\theta;h}$ is understood to be the maximal family of densities that can be expressed as (3.75) for some $\lambda \in \mathbb{R}^n$.

Joint h-Projection. Suppose a sample y^{N+m}, u^{N+m} is given with the empirical density $r_N(y,z)$. The necessary condition for $\hat{\lambda}$ to minimize the unnormalized joint inaccuracy (to maximize likelihood) is

$$0 = \nabla_\lambda K(r_N : s_{\theta,\hat{\lambda}})$$

$$= \iint r_N(y,z)\left(-h(y,z) + \iint s_{\theta,\hat{\lambda}}(y,z) h(y,z)\,dy\,dz\right) dy\,dz$$

$$= -\iint r_N(y,z) h(y,z)\,dy\,dz + \iint s_{\theta,\hat{\lambda}}(y,z) h(y,z)\,dy\,dz.$$

The condition thus reads

$$\boxed{\iint s_{\theta,\hat{\lambda}}(y,z) h(y,z)\,dy\,dz = \iint r_N(y,z) h(y,z)\,dy\,dz.} \tag{3.77}$$

The density $s_{\theta,\hat{\lambda}}(y,z)$ that satisfies the condition (3.77) will be called a *h-projection* of $r_N(y,z)$ onto $S_{\theta;h}$. Introducing the notation

$$\bar{h}_N \triangleq \iint r_N(y,z) h(y,z)\,dy\,dz = \frac{1}{N} \sum_{k=m+1}^{N+m} h(y_k, z_k), \tag{3.78}$$

$$\hat{h}(\theta, \lambda) \triangleq \iint s_{\theta,\lambda}(y,z) h(y,z)\,dy\,dz, \tag{3.79}$$

we can write (3.77) as

$$\hat{h}(\theta, \hat{\lambda}) = \bar{h}_N.$$

We denote the set of all densities $r(y, z)$ with the same h-projection as

$$\mathcal{R}_N \triangleq \left\{ r(y, z) : \iint r(y, z) h(y, z) \, dy \, dz = \bar{h}_N, \iint r(y, z) \, dy \, dz = 1, \, r(y, z) \geq 0 \right\}.$$

The expectation $\hat{h}(\theta, \lambda)$ can be viewed as an alternative way of parametrizing the joint exponential family $\mathcal{S}_{\theta;h}$ which is dual to the canonical λ-parametrization. The connection between both is exhibited by the fact that $\hat{h}(\theta, \lambda)$ coincides with the gradient of the normalizing divisor $\psi(\theta, \lambda)$ with respect to λ

$$
\begin{aligned}
\nabla_\lambda \psi(\theta, \lambda) &= \nabla_\lambda \log \iint s_\theta(y|z) \exp(\lambda^T h(y, z)) \, dy \, dz \\
&= \frac{\iint s_\theta(y|z) \nabla_\lambda \exp(\lambda^T h(y, z)) \, dy \, dz}{\iint s_\theta(y|z) \exp(\lambda^T h(y, z)) \, dy \, dz} \\
&= \iint \frac{s_\theta(y|z) \exp(\lambda^T h(y, z))}{\iint s_\theta(y|z) \exp(\lambda^T h(y, z)) \, dy \, dz} h(y, z) \, dy \, dz \\
&= \iint s_{\theta, \lambda}(y, z) h(y, z) \, dy \, dz \\
&= \hat{h}(\theta, \lambda).
\end{aligned}
\tag{3.80}
$$

Joint Pythagorean Relationship. Let $s_{\theta, \lambda}(y, z)$ be exponential (3.75) and $\hat{\lambda}$ satisfy (3.77). Then we can write

$$
\begin{aligned}
K(r_N &: s_\theta) - K(r_N : s_{\theta, \hat{\lambda}}) \\
&= \iint r_N(y, z) \log \frac{s_{\theta, \hat{\lambda}}(y, z)}{s_\theta(y|z)} \, dy \, dz \\
&= \iint r_N(y, z) \log \frac{s_\theta(y|z) \exp(\hat{\lambda}^T h(y, z) - \psi(\theta, \hat{\lambda}))}{s_\theta(y|z)} \, dy \, dz \\
&= \hat{\lambda}^T \left(\iint r_N(y, z) h(y, z) \, dy \, dz \right) - \psi(\theta, \hat{\lambda}) \\
&= \hat{\lambda}^T \left(\iint s_{\theta, \hat{\lambda}}(y, z) h(y, z) \, dy \, dz \right) - \psi(\theta, \hat{\lambda}) \\
&= \iint s_{\theta, \hat{\lambda}}(y, z) \log \frac{s_\theta(y|z) \exp(\hat{\lambda}^T h(y, z) - \psi(\theta, \hat{\lambda}))}{s_\theta(y|z)} \, dy \, dz \\
&= \iint s_{\theta, \hat{\lambda}}(y, z) \log \frac{s_{\theta, \hat{\lambda}}(y, z)}{s_\theta(y|z)} \, dy \, dz \\
&= D(s_{\theta, \hat{\lambda}} \| s_\theta)
\end{aligned}
$$

where we used the notation $D(s_{\theta, \hat{\lambda}} \| s_\theta)$ for the *unnormalized* joint K-L distance. In such a way, we obtain an unnormalized joint analogue of the Pythagorean relationship (cf. Fig. 3.11)

$$K(r_N : s_\theta) = K(r_N : s_{\theta,\hat{\lambda}}) + D(s_{\theta,\hat{\lambda}} \| s_\theta). \tag{3.81}$$

Let us stress that in contrast to (3.31), both $K(r_N : s_\theta)$ and $D(s_{\theta,\hat{\lambda}} \| s_\theta)$ are taken now with respect to the function

$$s_\theta(y|z)\, 1(z)$$

which is not normalized, i.e., $\iint s_\theta(y|z)\, dy\, dz \neq 1$ in general. This, of course, affects to some degree the meaning and properties of the measures. In particular, the "K-L distance" $D(s_{\theta,\hat{\lambda}} \| s_\theta)$ is not bounded from below by zero any more. At any rate, the Pythagorean relationship (3.81) still enables us to decompose $K(r_N : s_\theta)$ into two terms, one of which can be made independent of θ by a proper choice of the canonical statistic $h(y,z)$. This is what matters to us at the moment. In Sect. 4.3, the measure $D(s_{\theta,\hat{\lambda}} \| s_\theta)$ is given a natural interpretation in approximate estimation.

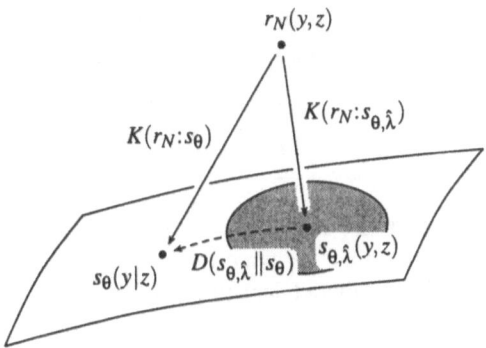

Fig. 3.11. Pythagorean-like decomposition of inaccuracy for dependent observations. The projection "surface" corresponds to the set of all functions $C s_\theta(y|z) \exp(\lambda^T h(y,z))$ with $C > 0$. The shaded area indicates a subset of normalized densities with $C = \exp(-\psi(\theta,\lambda))$.

Minimum Joint Inaccuracy Projection. Assume that the joint inaccuracy $K(r_N : s_{\theta,\hat{\lambda}})$ of $r_N(y,z)$ relative to the h-projection $s_{\theta,\hat{\lambda}}(y,z)$ is finite

$$K(r_N : s_{\theta,\hat{\lambda}}) < \infty. \tag{3.82}$$

The following Pythagorean relationship holds for every $\lambda \in \mathcal{N}_\theta$

$$
\begin{aligned}
& K(r_N : s_{\theta,\lambda}) - K(r_N : s_{\theta,\hat{\lambda}}) \\
&= \iint r_N(y,z) \log \frac{s_{\theta,\hat{\lambda}}(y,z)}{s_{\theta,\lambda}(y,z)}\, dy\, dz \\
&= \iint r_N(y,z) \log \frac{s_\theta(y|z) \exp(\hat{\lambda}^T h(y,z) - \psi(\theta,\hat{\lambda}))}{s_\theta(y|z) \exp(\lambda^T h(y,z) - \psi(\theta,\lambda))}\, dy\, dz
\end{aligned}
$$

$$= (\hat{\lambda}-\lambda)^T \left(\iint r_N(y,z)\, h(y,z)\, dy\, dz \right) - \psi(\theta,\hat{\lambda}) + \psi(\theta,\lambda)$$

$$= (\hat{\lambda}-\lambda)^T \left(\iint s_{\theta,\hat{\lambda}}(y,z)\, h(y,z)\, dy\, dz \right) - \psi(\theta,\hat{\lambda}) + \psi(\theta,\lambda)$$

$$= \iint s_{\theta,\hat{\lambda}}(y,z) \log \frac{s_\theta(y|z)\, \exp(\hat{\lambda}^T h(y,z) - \psi(\theta,\hat{\lambda}))}{s_\theta(y|z)\, \exp(\lambda^T h(y,z) - \psi(\theta,\lambda))}\, dy\, dz$$

$$= \iint s_{\theta,\hat{\lambda}}(y,z) \log \frac{s_{\theta,\hat{\lambda}}(y,z)}{s_{\theta,\lambda}(y,z)}\, dy\, dz$$

$$= D(s_{\theta,\hat{\lambda}} \| s_{\theta,\lambda}).$$

Since the joint K-L distance $D(s_{\theta,\hat{\lambda}} \| s_{\theta,\lambda})$ is nonnegative by (3.16) and the inaccuracy $K(r_N{:}s_{\theta,\hat{\lambda}})$ was assumed finite in (3.82), we have

$$K(r_N{:}s_{\theta,\lambda}) \ge K(r_N{:}s_{\theta,\hat{\lambda}})$$

with equality if and only if $s_{\theta,\lambda}(y,z) = s_{\theta,\hat{\lambda}}(y,z)$ almost everywhere. Thus, under the assumption (3.82), the h-projection $s_{\theta,\hat{\lambda}}(y,z)$ is a unique solution to the minimum inaccuracy problem

$$\boxed{K(r_N{:}s_{\theta,\hat{\lambda}}) = \min_{\lambda \in \mathcal{N}_\theta} K(r_N{:}s_{\theta,\lambda}).} \tag{3.83}$$

Minimum Joint K-L Distance Projection. A dual interpretation of the h-projection looks as follows. Assume that the unnormalized K-L distance of the h-projection $s_{\theta,\hat{\lambda}}(y,z)$ and the conditional model density $s_\theta(y|z)$ is finite

$$D(s_{\theta,\hat{\lambda}} \| s_\theta) < \infty. \tag{3.84}$$

Then there are $r(y,z) \in \mathcal{R}_N$ such that $D(r\|s_\theta) < \infty$. For every such $r(y,z) \in \mathcal{R}_N$, the following Pythagorean relation holds

$$D(r\|s_\theta) - D(r\|s_{\theta,\hat{\lambda}})$$

$$= \iint r(y,z) \log \frac{s_{\theta,\hat{\lambda}}(y,z)}{s_\theta(y|z)}\, dy\, dz$$

$$= \iint r(y,z) \log \frac{s_\theta(y|z)\, \exp(\hat{\lambda}^T h(y,z) - \psi(\theta,\hat{\lambda}))}{s_\theta(y|z)}\, dy\, dz$$

$$= \hat{\lambda}^T \left(\iint r(y,z)\, h(y,z)\, dy\, dz \right) - \psi(\theta,\hat{\lambda})$$

$$= \hat{\lambda}^T \left(\iint s_{\theta,\hat{\lambda}}(y,z)\, h(y,z)\, dy\, dz \right) - \psi(\theta,\hat{\lambda})$$

$$= \iint s_{\theta,\hat{\lambda}}(y,z) \log \frac{s_\theta(y|z)\, \exp(\hat{\lambda}^T h(y,z) - \psi(\theta,\hat{\lambda}))}{s_\theta(y|z)}\, dy\, dz$$

$$= \iint s_{\theta,\hat{\lambda}}(y,z) \log \frac{s_{\theta,\hat{\lambda}}(y,z)}{s_\theta(y|z)}\, dy\, dz$$

$$= D(s_{\theta,\hat{\lambda}} \| s_\theta).$$

Since the joint K-L distance $D(r\|s_{\theta,\hat{\lambda}})$ is nonnegative by (3.16) and we consider only $r(y,z) \in \mathcal{R}_N$ such that $D(r\|s_\theta) < \infty$, we have for every such $r(y,z)$

$$D(r\|s_\theta) \geq D(s_{\theta,\hat{\lambda}}\|s_\theta)$$

with equality if and only if $r(y,z) = s_{\theta,\hat{\lambda}}(y,z)$ almost everywhere. Thus, under the assumption (3.84), the h-projection $s_{\theta,\hat{\lambda}}(y,z)$ is a unique solution to the minimum K-L distance problem

$$D(s_{\theta,\hat{\lambda}}\|s_\theta) = \min_{r\in\mathcal{R}_N} D(r\|s_\theta). \qquad (3.85)$$

Minimum Joint K-L Distance. One possible way of calculating the minimum unnormalized K-L distance $D(s_{\theta,\hat{\lambda}}\|s_\theta)$ is to determine the h-projection $s_{\theta,\hat{\lambda}}(y,z)$ explicitly and substitute it in $D(s_{\theta,\hat{\lambda}}\|s_\theta)$. This yields

$$\begin{aligned}
D(s_{\theta,\hat{\lambda}}\|s_\theta) &= \iint s_{\theta,\hat{\lambda}}(y,z) \log \frac{s_{\theta,\hat{\lambda}}(y,z)}{s_\theta(y|z)}\,dy\,dz \\
&= \iint s_{\theta,\hat{\lambda}}(y,z) \log \frac{s_\theta(y|z)\exp(\hat{\lambda}^T h(y,z) - \psi(\theta,\hat{\lambda}))}{s_\theta(y|z)}\,dy\,dz \\
&= \hat{\lambda}^T \hat{h}(\theta,\hat{\lambda}) - \psi(\theta,\hat{\lambda}) \\
&= \hat{\lambda}^T \bar{h}_N - \psi(\theta,\hat{\lambda}).
\end{aligned} \qquad (3.86)$$

Another possibility is

$$\begin{aligned}
0 &= \min_\lambda D(s_{\theta,\hat{\lambda}}\|s_{\theta,\lambda}) \\
&= \min_\lambda \iint s_{\theta,\hat{\lambda}}(y,z) \log \frac{s_{\theta,\hat{\lambda}}(y,z)}{s_\theta(y|z)\exp(\lambda^T h(y,z) - \psi(\theta,\lambda))}\,dy\,dz \\
&= D(s_{\theta,\hat{\lambda}}\|s_\theta) - \max_\lambda \left(\lambda^T \hat{h}(\theta,\hat{\lambda}) - \psi(\theta,\lambda)\right) \\
&= D(s_{\theta,\hat{\lambda}}\|s_\theta) - \max_\lambda \left(\lambda^T \bar{h}_N - \psi(\theta,\lambda)\right).
\end{aligned}$$

As a result, we have the expression

$$D(s_{\theta,\hat{\lambda}}\|s_\theta) = \max_\lambda \left(\lambda^T \bar{h}_N - \psi(\theta,\lambda)\right). \qquad (3.87)$$

Thus, the minimum K-L distance follows by maximizing $\lambda^T \bar{h}_N - \psi(\theta,\lambda)$ over λ.

Enveloping Joint Exponential Family. Taking (3.81) and (3.87) together, we have

$$K(r_N:s_\theta) = K(r_N:s_{\theta,\hat{\lambda}}) + \max_\lambda \left(\lambda^T \bar{h}_N - \psi(\theta,\lambda)\right). \qquad (3.88)$$

The Pythagorean relationship (3.81) enables us to evaluate the inaccuracy $K(r_N:s_\theta)$, with precision up to an additive constant, *without* knowledge of $r_N(y,z)$ *provided* $K(r_N:s_{\theta,\hat{\lambda}})$ is independent of θ for every $r_N(y,z)$.

The latter means that for every θ, θ' and every $r_N(y,z)$, it holds

$$K(r_N{:}s_{\theta,\hat{\lambda}}) = K(r_N{:}s_{\theta',\hat{\lambda}'})$$

where $\hat{\lambda}$, $\hat{\lambda}'$ are such that $\hat{h}(\theta,\hat{\lambda}) = \hat{h}(\theta',\hat{\lambda}') = \bar{h}_N$. This condition is to be satisfied for every $r_N(y,z)$, including

$$r_N(y,z) = \delta(y-a, z-b), \quad a \in \mathcal{Y}, \quad b \in \mathcal{Z}$$

which implies

$$s_{\theta,\hat{\lambda}}(y,z) = s_{\theta',\hat{\lambda}'}(y,z).$$

Hence, the h-projections of any $r_N(y,z)$ onto $\mathcal{S}_{\theta;h}$ and $\mathcal{S}_{\theta';h}$ coincide. But this may happen only if the exponential families $\mathcal{S}_{\theta;h}$ and $\mathcal{S}_{\theta';h}$ coincide as whole. If this is the case for every $\theta \in \mathcal{T}$, the model family $\mathcal{S}$ can be parametrized so that for every θ there exists $\lambda(\theta)$ such that

$$s_\theta(y|z) = C s_{\theta_0}(y|z) \exp\!\left(\lambda^T(\theta) h(y,z)\right) \tag{3.89}$$

where $s_{\theta_0}(y|z)$ is a fixed conditional density from the model family $\mathcal{S}$ and C is a constant independent of (y,z).

Thus, the unnormalized joint inaccuracy $K(\bar{r}_N{:}s_{\theta,\hat{\lambda}})$ is independent of θ provided the function $s_\theta(y|z)$ for every θ belongs to an unnormalized exponential family of positive functions

$$\left\{ C s_{\theta_0}(y|z) \exp\!\left(\lambda^T h(y,z)\right) : C > 0, \, \lambda \in \mathbb{R}^n \right\}$$

with a fixed origin $s_{\theta_0}(y|z)$ and the canonical statistic $h(y,z)$. Therefore, if we choose $h(y,z)$ as a canonical statistic of any unnormalized exponential family *enveloping* the model family $\mathcal{S}$, then (3.89) is satisfied by definition and $K(\bar{r}_N{:}s_{\theta,\hat{\lambda}})$ is independent of θ.

Under the condition (3.89), it follows from (3.88) that

$$\boxed{K(r_N{:}s_\theta) = C + \max_\lambda \left(\lambda^T \bar{h}_N - \psi(\theta,\lambda)\right)} \tag{3.90}$$

where C is a constant independent of θ.

Equivalence of Conditional and Joint Views. Assume that the conditional exponential family $\bar{\mathcal{S}}_{\theta;f}$ is independent of θ, i.e., given any two θ, θ', for every τ there exists τ' such that $s_{\theta,\tau}(y|z) = s_{\theta',\tau'}(y|z)$ and, consequently,

$$\frac{s_\theta(y|z) \exp\!\left(\tau^T f(y;z)\right)}{s_{\theta'}(y|z) \exp\!\left(\tau'^T f(y;z)\right)} = \frac{\int s_\theta(y|z) \exp\!\left(\tau^T f(y;z)\right) dy}{\int s_{\theta'}(y|z) \exp\!\left(\tau'^T f(y;z)\right) dy}. \tag{3.91}$$

In addition, assume that the marginal exponential family $\mathcal{W}_{\theta,\tau;g}$ is independent of (θ,τ), i.e., given any two (θ,τ) and (θ',τ'), for every μ there exists μ' such that $w_{\theta,\tau,\mu}(z) = w_{\theta',\tau',\mu'}(z)$ and, consequently,

$$\frac{\int s_\theta(y|z) \exp\left(\tau^T f(y;z) + \mu^T g(z)\right) dy}{\int s_{\theta'}(y|z) \exp\left(\tau'^T f(y;z) + \mu'^T g(z)\right) dy}$$

$$= \frac{\iint s_\theta(y|z) \exp\left(\tau^T f(y;z) + \mu^T g(z)\right) dy\, dz}{\iint s_{\theta'}(y|z) \exp\left(\tau'^T f(y;z) + \mu'^T g(z)\right) dy\, dz}. \tag{3.92}$$

Taking (3.91) and (3.92) together, we obtain that given any two θ, θ', for every (τ, μ) there exists (τ', μ') such that

$$\frac{s_\theta(y|z) \exp\left(\tau^T f(y;z) + \mu^T g(z)\right)}{s_{\theta'}(y|z) \exp\left(\tau'^T f(y;z) + \mu'^T g(z)\right)}$$

$$= \frac{\iint s_\theta(y|z) \exp\left(\tau^T f(y;z) + \mu^T g(z)\right) dy\, dz}{\iint s_{\theta'}(y|z) \exp\left(\tau'^T f(y;z) + \mu'^T g(z)\right) dy\, dz}, \tag{3.93}$$

i.e., $s_{\theta,\tau,\mu}(y,z) = s_{\theta',\tau',\mu'}(y,z)$. Hence, the joint exponential family $S_{\theta;g,h}$ is independent of θ.

Conversely, assume that the joint exponential family $S_{\theta,\lambda;h}$ is independent of θ, i.e., given any two θ and θ', for every λ there exists λ' such that $s_{\theta,\lambda}(y,z) = s_{\theta',\lambda'}(y,z)$ and, consequently,

$$\frac{s_\theta(y|z) \exp\left(\lambda^T h(y,z)\right)}{s_{\theta'}(y|z) \exp\left(\lambda'^T h(y,z)\right)} = \frac{\iint s_\theta(y|z) \exp\left(\lambda^T h(y,z)\right) dy\, dz}{\iint s_{\theta'}(y|z) \exp\left(\lambda'^T h(y,z)\right) dy\, dz}. \tag{3.94}$$

Now suppose there exists a set of functions $g_0(z) \equiv 1, g_1(z), \dots, g_{n''}(z)$ such that the linear space spanned by these functions is a subspace of the linear space spanned by the functions $h_0(y,z) \equiv 1, h_1(y,z), \dots, h_n(y,z)$

$$\mathcal{G} \overset{\triangle}{=} \text{span}\left\{g_0(z), g_1(z), \dots, g_{n''}(z)\right\} \subset \mathcal{H} = \text{span}\left\{h_0(y,z), h_1(y,z), \dots, h_n(y,z)\right\}.$$

Consider the quotient space $\mathcal{H}/\mathcal{G}$ which is composed of equivalence classes of those functions $h(y,z)$ that differ by a function from $\mathcal{G}$ only, i.e., $h(y,z)$ and $h'(y,z)$ are taken as equivalent if $h(y,z) - h'(y,z) \in \mathcal{G}$. The quotient space $\mathcal{H}/\mathcal{G}$ of the linear space $\mathcal{H}$ is a linear space again with operations defined through arbitrary representatives of the equivalence classes. If $\mathcal{H}$ is of dimension n and $\mathcal{G}$ is of dimension n'', then $\mathcal{H}/\mathcal{G}$ is of dimension $n' = n - n''$. We choose one representative from each class

$$f_1(y;z), \dots, f_{n'}(y;z).$$

Rewriting (3.94) in terms of the statistics $f(y;z)$ and $g(z)$, we obtain (3.93). The identity (3.93) clearly implies (3.92), i.e., the marginal exponential family $W_{\theta,\tau;g}$ is independent of (θ, τ). The identities (3.93) and together (3.92) imply (3.91), i.e., the conditional exponential family $\tilde{S}_{\theta;f}$ is independent of θ.

More straightforwardly perhaps, the equivalence of the conditional and joint views of parameter estimation can be seen directly by comparing the conditions (3.56) and (3.69) with (3.89).

With respect to the equivalence of the above views, we use in the following only the latter, more compact view via joint densities $s_{\theta,\lambda}(y,z)$.

Example 3.5 (*Normal distribution*) Consider the normal conditional sampling density

$$s_\theta(y|z) = \frac{1}{\sqrt{2\pi}} \exp\left(-\frac{1}{2}(y - \theta z)^2\right).$$

The density clearly admits

$$s_\theta(y|z) = C s_{\theta_0}(y|z) \exp\left(\lambda_1 h_1(y,z) + \lambda_2 h_2(y,z)\right)$$

where C is a constant independent of θ, y, z and

$$h_1(y,z) = yz + z^2,$$
$$h_2(y,z) = yz - z^2.$$

The linear space spanned by the above functions is composed of linear combinations

$$\lambda_1 (yz + z^2) + \lambda_2 (yz - z^2).$$

The linear combinations of $h_1(y,z)$ and $h_2(y,z)$ that depend only on z can be found by solving the equation

$$\frac{d}{dy}\left(\lambda_1 (yz + z^2) + \lambda_2 (yz - z^2)\right) = (\lambda_1 + \lambda_2) z = 0$$

with respect to λ_1 and λ_2. Choosing $\lambda_1 = -\lambda_2 = 1/2$, we get $g(z) = z^2$. The statistic $f(y;z)$ follows by the requirement of linear independence of $g(z)$ and $f(y;z)$. This is satisfied, for instance, by the choice $f(y;z) = h_1(y,z) = yz - z^2$. □

3.4 Examples: Normal and Linear Models

To give the reader a better feel for the Pythagorean-like estimation via (3.41) and (3.90), we apply the general results shown above to models with normal stochastics and linear dynamics. Not surprisingly, the results obtained in this way coincide with the standard solutions. The major merit of the Pythagorean approach is not, of course, in deriving old results in a new way but in its potential to be extended quite straightforwardly to the models with non-normal stochastics and non-linear dynamics. This will be illustrated later—in Chap. 4.

Independent Observations

Consider the case of normally distributed data $Y_k \sim N(\theta, 1)$ with the sampling density

$$s_\theta(y) = \frac{1}{\sqrt{2\pi}} \exp\left(-\frac{1}{2}(y - \theta)^2\right).$$

Given an arbitrary fixed $\theta \in \mathcal{T}$, we construct the enveloping exponential family $S_{\theta;h}$ composed of densities

$$s_{\theta,\lambda}(y) = s_\theta(y)\exp\big(\lambda h(y) - \psi(\theta,\lambda)\big)$$

with $h(y) = y$ and

$$
\begin{aligned}
\psi(\theta,\lambda) &= \log \int s_\theta(y)\exp(\lambda h(y))\,dy \\
&= \log \int \frac{1}{\sqrt{2\pi}}\exp\Big(-\frac{1}{2}(y-\theta)^2\Big)\exp(\lambda y)\,dy \\
&= \log \int \frac{1}{\sqrt{2\pi}}\exp\Big(-\frac{1}{2}(y-(\theta+\lambda))^2\Big)\,dy \\
&\quad -\frac{1}{2}\big(\theta^2 - (\theta+\lambda)^2\big) \\
&= \theta\lambda + \frac{1}{2}\lambda^2.
\end{aligned}
$$

After substituting for $s_\theta(y)$ and some simple algebraic manipulations, we obtain

$$s_{\theta,\lambda}(y) = \frac{1}{\sqrt{2\pi}}\exp\Big(-\frac{1}{2}(y-(\theta+\lambda))^2\Big).$$

It is easy to see that the family $S_{\theta;h}$ is independent of θ. Indeed, given any θ, θ', we can assign to every λ the value

$$\lambda' = \lambda + \theta - \theta'$$

so that

$$s_{\theta,\lambda}(y) = s_{\theta',\lambda'}(y).$$

Hence, we can apply the formula (3.41)

$$
\begin{aligned}
K(r_N{:}s_\theta) &= C + \max_\lambda\big(\lambda\bar{h}_N - \psi(\theta,\lambda)\big) \\
&= C + \max_\lambda\Big(\lambda E_N(Y) - \theta\lambda - \frac{1}{2}\lambda^2\Big).
\end{aligned}
$$

The value $\hat{\lambda}$ that maximizes the concave function

$$\lambda E_N(Y) - \theta\lambda - \frac{1}{2}\lambda^2$$

follows from the condition

$$\frac{d}{d\lambda}\Big(\lambda E_N(Y) - \theta\lambda - \frac{1}{2}\lambda^2\Big)\Big|_{\lambda=\hat{\lambda}} = E_N(Y) - \theta - \hat{\lambda} = 0.$$

Given the solution $\hat{\lambda} = E_N(Y) - \theta$, we obtain

$$
\begin{aligned}
K(r_N{:}s_\theta) &= C + \hat{\lambda}E_N(Y) - \theta\hat{\lambda} - \frac{1}{2}\hat{\lambda}^2 \\
&= C + \frac{1}{2}\big(\theta - E_N(Y)\big)^2.
\end{aligned}
$$

Finally, substituting the last expression for $K(r_N{:}s_\theta)$ in (2.23), we obtain the likelihood function

$$l_N(\theta) = C'\exp\Big(-\frac{1}{2}N(\theta - E_N(Y))^2\Big).$$

It clearly coincides with the standard solution.

Controlled Dynamic Systems:
View via Conditional Densities

Consider the first-order linear normal autoregression $Y_k \sim N(\theta Y_{k-1}, 1)$ that implies the conditional sampling density of the form

$$s_\theta(y|z) = \frac{1}{\sqrt{2\pi}} \exp\left(-\frac{1}{2}(y - \theta z)^2\right)$$

with $z_k = y_{k-1}$.

Given an arbitrary fixed $\theta \in \mathcal{T}$, we construct an enveloping exponential family $\bar{s}_{\theta;f}$ composed of the conditional densities

$$s_{\theta,\tau}(y|z) = s_\theta(y|z) \exp\left(\tau f(y;z) - \psi(\theta, \tau; z)\right)$$

with $f(y;z) = yz$ and

$$
\begin{aligned}
\psi(\theta, \tau; z) &= \log \int s_\theta(y|z) \exp(\tau f(y;z)) \, dy \\
&= \log \int \frac{1}{\sqrt{2\pi}} \exp\left(-\frac{1}{2}(y - \theta z)^2\right) \exp(\tau yz) \, dy \\
&= \log \int \frac{1}{\sqrt{2\pi}} \exp\left(-\frac{1}{2}(y - (\theta + \tau)z)^2\right) dy - \frac{1}{2}(\theta^2 - (\theta + \tau)^2) z^2 \\
&= \frac{1}{2}(2\theta\tau + \tau^2) z^2.
\end{aligned}
$$

After substituting for $s_\theta(y|z)$ and some algebraic manipulations, we find

$$s_{\theta,\tau}(y|z) = \frac{1}{\sqrt{2\pi}} \exp\left(-\frac{1}{2}(y - (\theta + \tau)z)^2\right).$$

The family $\bar{s}_{\theta;f}$ is independent of θ as given any θ, θ', we can assign to every τ the value

$$\tau' = \tau + \theta - \theta'$$

so that

$$s_{\theta,\tau}(y|z) = s_{\theta',\tau'}(y|z).$$

Hence, we can use the formula (3.57) to obtain

$$
\begin{aligned}
\bar{K}(r_N : s_\theta) &= C + \max_\tau \left(\tau \bar{f}_N - E_N\left(\psi(\theta, \tau; Z)\right)\right) \\
&= C + \max_\tau \left(\tau E_N(YZ) - E_N\left(\psi(\theta, \tau; Z)\right)\right). \tag{3.95}
\end{aligned}
$$

The empirical expectation above can be written as the unnormalized inaccuracy

$$E_N\left(-\psi(\theta, \tau; Z)\right) = K(\bar{r}_N : w_{\theta,\tau}) \tag{3.96}$$

relative to the function

$$w_{\theta,\tau}(z) = \exp\left(-\frac{1}{2}\left(-\tau^2 - 2\theta\tau\right)z^2\right).$$

Note that $w_{\theta,\tau}(z)$ is not normalized, hence it is not a probability density function.

We construct the enveloping exponential family $\mathcal{W}_{\theta,\tau;g}$ composed of the marginal densities

$$w_{\theta,\tau,\mu}(z) = w_{\theta,\tau}(z)\exp\left(\mu g(z) - \psi(\theta,\tau,\mu)\right)$$

with $g(z) = z^2$ and

$$
\begin{aligned}
\psi(\theta,\tau,\mu) &= \log\int w_{\theta,\tau}(z)\exp\left(\mu g(z)\right)dz \\
&= \log\int \exp\left(-\frac{1}{2}\left(-\tau^2 - 2\theta\tau - 2\mu\right)z^2\right)dz \\
&= \frac{1}{2}\log 2\pi - \frac{1}{2}\log\left(-\tau^2 - 2\theta\tau - 2\mu\right).
\end{aligned}
$$

After substituting for $w_{\theta,\tau}(z)$, we have

$$w_{\theta,\tau,\mu}(z) = \frac{1}{\sqrt{2\pi}}\left(-\tau^2 - 2\theta\tau - 2\mu\right)^{\frac{1}{2}}\exp\left(-\frac{1}{2}\left(-\tau^2 - 2\theta\tau - 2\mu\right)z^2\right).$$

The family $\mathcal{W}_{\theta,\tau;g}$ is clearly independent of (θ,τ) because given any (θ,τ), (θ',τ'), one can assign to every μ the value

$$\mu' = \mu + \frac{1}{2}\tau^2 + \theta\tau - \frac{1}{2}\tau'^2 - \theta'\tau'$$

so that

$$w_{\theta,\tau,\mu}(z) = w_{\theta',\tau',\mu'}(z).$$

Hence, we can use the formula (3.70)

$$
\begin{aligned}
K(\bar{r}_N : w_{\theta,\tau}) &= \max_{\mu}\left(\mu\bar{g}_N - \psi(\theta,\tau,\mu)\right) \\
&= \max_{\mu}\left(\mu E_N(Z^2) - \frac{1}{2}\log 2\pi + \frac{1}{2}\log\left(-\tau^2 - 2\theta\tau - 2\mu\right)\right). \quad (3.97)
\end{aligned}
$$

Putting (3.95), (3.96) and (3.97) together, or equivalently, using (3.71), we obtain the formula

$$\bar{K}(r_N : s_\theta) = C + \max_{\tau,\mu}\left(\tau E_N(YZ) + \mu E_N(Z^2) - \frac{1}{2}\log 2\pi + \frac{1}{2}\log\left(-\tau^2 - 2\theta\tau - 2\mu\right)\right)$$

where C is a constant independent of θ

The values $(\hat{\tau}, \hat{\mu})$ that maximize the concave function

$$\tau E_N(YZ) + \mu E_N(Z^2) - \frac{1}{2}\log 2\pi + \frac{1}{2}\log\left(-\tau^2 - 2\theta\tau - 2\mu\right)$$

can be found from the conditions

$$\frac{\partial}{\partial \tau}\left(\tau E_N(YZ) + \mu E_N(Z^2) - \frac{1}{2}\log 2\pi + \frac{1}{2}\log(-\tau^2 - 2\theta\tau - 2\mu)\right)\Big|_{\tau = \hat{\tau}, \mu = \hat{\mu}}$$

$$= E_N(YZ) + \frac{1}{2}\frac{-2\theta - 2\hat{\tau}}{-\hat{\tau}^2 - 2\theta\hat{\tau} - 2\hat{\mu}} = 0,$$

$$\frac{\partial}{\partial \mu}\left(\tau E_N(YZ) + \mu E_N(Z^2) - \frac{1}{2}\log 2\pi + \frac{1}{2}\log(-\tau^2 - 2\theta\tau - 2\mu)\right)\Big|_{\tau = \hat{\tau}, \mu = \hat{\mu}}$$

$$= E_N(Z^2) + \frac{1}{2}\frac{-2}{-\hat{\tau}^2 - 2\theta\hat{\tau} - 2\hat{\mu}} = 0.$$

After a bit of computing, we find that the maximum is attained at

$$\hat{\tau} = \hat{\theta}_N - \theta,$$

$$\hat{\mu} = \frac{1}{2}\left(\theta^2 - \hat{\theta}_N^2\right) - \frac{1}{2}\frac{1}{E_N(Z^2)}$$

where we introduced (for reasons that become clear at once) the notation

$$\hat{\theta}_N = \frac{E_N(YZ)}{E_N(Z^2)}.$$

The corresponding maximum value is obtained after necessary algebraic manipulations

$$\bar{K}(r_N : s_\theta) = C + \hat{\tau} E_N(YZ) + \hat{\mu} E_N(Z^2) - \frac{1}{2}\log 2\pi + \frac{1}{2}\log(-\hat{\tau}^2 - 2\theta\hat{\tau} - 2\hat{\mu})$$

$$= C' + \frac{1}{2}E_N(Z^2)(\theta - \hat{\theta}_N)^2.$$

Finally, substituting the last expression for $\bar{K}(r_N : s_\theta)$ in (2.33), we obtain the likelihood function

$$l_N(\theta) = C'' \exp\left(-\frac{1}{2}N E_N(Z^2)(\theta - \hat{\theta}_N)^2\right)$$

which clearly coincides with the standard solution.

Controlled Dynamic Systems:
View via Joint Densities

Once again, we consider the first-order linear normal autoregression $Y_k \sim N(\theta Y_{k-1}, 1)$ that implies the conditional sampling density of the form

$$s_\theta(y|z) = \frac{1}{\sqrt{2\pi}} \exp\left(-\frac{1}{2}(y - \theta z)^2\right)$$

where $z_k = y_{k-1}$.

In contrast to the preceding case, we construct an enveloping exponential family $S_{\theta;h}$ composed of the *joint* densities

$$s_{\theta,\lambda}(y, z) = s_\theta(y|z) \exp\left(\lambda^T h(y, z) - \psi(\theta, \lambda)\right)$$

with $h_1(y,z) = yz$, $h_2(y,z) = z^2$ and

$$
\begin{aligned}
\psi(\theta,\lambda) &= \log \iint s_\theta(y|z) \exp(\lambda^T h(y,z)) \, dy \, dz \\
&= \log \iint \frac{1}{\sqrt{2\pi}} \exp\left(-\frac{1}{2}(y-\theta z)^2\right) \exp(\lambda_1 yz + \lambda_2 z^2) \, dy \, dz \\
&= \log \iint \frac{1}{\sqrt{2\pi}} \exp\left(-\frac{1}{2} \begin{bmatrix} y \\ z \end{bmatrix}^T \begin{bmatrix} 1 & -\theta-\lambda_1 \\ -\theta-\lambda_1 & \theta^2-2\lambda_2 \end{bmatrix} \begin{bmatrix} y \\ z \end{bmatrix}\right) dy \, dz \\
&= \log \sqrt{2\pi} \left| \begin{matrix} 1 & -\theta-\lambda_1 \\ -\theta-\lambda_1 & \theta^2-2\lambda_2 \end{matrix} \right|^{-\frac{1}{2}} \\
&= \frac{1}{2}\log 2\pi - \frac{1}{2}\log(-\lambda_1^2 - 2\theta\lambda_1 - 2\lambda_2).
\end{aligned}
$$

After substituting for $s_\theta(y|z)$ and some manipulations, we obtain

$$
s_{\theta,\lambda}(y,z) = \frac{1}{2\pi} \left| \begin{matrix} 1 & -\theta-\lambda_1 \\ -\theta-\lambda_1 & \theta^2-2\lambda_2 \end{matrix} \right|^{\frac{1}{2}} \exp\left(-\frac{1}{2} \begin{bmatrix} y \\ z \end{bmatrix}^T \begin{bmatrix} 1 & -\theta-\lambda_1 \\ -\theta-\lambda_1 & \theta^2-2\lambda_2 \end{bmatrix} \begin{bmatrix} y \\ z \end{bmatrix}\right).
$$

The family $S_{\theta;h}$ is independent of θ as given any θ, θ', we can assign to every (λ_1, λ_2) the values

$$
\begin{aligned}
\lambda_1' &= \lambda_1 + \theta - \theta' \\
\lambda_2' &= \lambda_2 - \frac{1}{2}\theta^2 + \frac{1}{2}\theta'^2
\end{aligned}
$$

so that

$$
s_{\theta,\lambda}(y,z) = s_{\theta',\lambda'}(y,z).
$$

Hence, we can use the formula (3.90) to obtain

$$
\begin{aligned}
K(r_N : s_\theta) &= C + \max_\lambda \left(\lambda \bar{h}_N - \psi(\theta,\lambda) \right) \\
&= \max_\lambda \left(\lambda_1 E_N(YZ) + \lambda_2 E_N(Z^2) - \frac{1}{2}\log 2\pi + \frac{1}{2}\log(-\lambda_1^2 - 2\theta\lambda_1 - 2\lambda_2) \right).
\end{aligned}
$$

The rest of computation is like in the preceding case. The values $\hat{\lambda}_1$ and $\hat{\lambda}_2$ that maximize the concave function

$$
\lambda_1 E_N(YZ) + \lambda_2 E_N(Z^2) - \frac{1}{2}\log 2\pi + \frac{1}{2}\log(-\lambda_1^2 - 2\theta\lambda_1 - 2\lambda_2)
$$

follow from the conditions

$$
\begin{aligned}
&\frac{\partial}{\partial \lambda_1}\left(\lambda_1 E_N(YZ) + \lambda_2 E_N(Z^2) - \frac{1}{2}\log 2\pi + \frac{1}{2}\log(-\lambda_1^2 - 2\theta\lambda_1 - 2\lambda_2) \right)\Big|_{\lambda=\hat{\lambda}} \\
&\qquad = E_N(YZ) + \frac{1}{2}\frac{-2\theta - 2\hat{\lambda}_1}{-\hat{\lambda}_1^2 - 2\theta\hat{\lambda}_1 - 2\hat{\lambda}_2} = 0, \\
&\frac{\partial}{\partial \lambda_2}\left(\lambda_2 E_N(YZ) + \lambda_2 E_N(Z^2) - \frac{1}{2}\log 2\pi + \frac{1}{2}\log(-\lambda_1^2 - 2\theta\lambda_1 - 2\lambda_2) \right)\Big|_{\lambda=\hat{\lambda}} \\
&\qquad = E_N(Z^2) + \frac{1}{2}\frac{-2}{-\hat{\lambda}_1^2 - 2\theta\hat{\lambda}_1 - 2\hat{\lambda}_2} = 0.
\end{aligned}
$$

After some computation, we obtain

$$\hat{\lambda}_1 = \hat{\theta}_N - \theta,$$
$$\hat{\lambda}_2 = \frac{1}{2}\left(\theta^2 - \hat{\theta}_N^2\right) - \frac{1}{2}\frac{1}{E_N(Z^2)}$$

with the notation

$$\hat{\theta}_N = \frac{E_N(YZ)}{E_N(Z^2)}.$$

The corresponding maximum value is found after straightforward algebraic manipulations

$$K(r_N{:}s_\theta) = C + \hat{\lambda}_1 E_N(YZ) + \hat{\lambda}_2 E_N(Z^2) - \frac{1}{2}\log 2\pi + \frac{1}{2}\log\left(-\hat{\lambda}_1^2 - 2\theta\hat{\lambda}_1 - 2\hat{\lambda}_2\right)$$
$$= C' + \frac{1}{2}(\theta - \hat{\theta}_N)^2 E_N(Z^2).$$

Finally, taking into account $\bar{K}(r_N{:}s_\theta) = K(r_N{:}s_\theta)$ and substituting the last expression for $\bar{K}(r_N{:}s_\theta)$ in (2.33), we obtain the likelihood function

$$l_N(\theta) = C'' \exp\left(-\frac{1}{2}N E_N(Z^2)\left(\theta - \hat{\theta}_N\right)^2\right)$$

again.

3.5 Data Compression

As we have pointed out earlier on several occasions, the Pythagorean relationship enables us in certain cases to restore the likelihood function of the posterior density without complete knowledge of the empirical density. This feature is analysed in detail now.

Independent Observations

Under the condition (3.40), all information which is required from data to reconstruct $K(r_N{:}s_\theta)$ through (3.41)

$$K(r_N{:}s_\theta) = C + \max_\lambda \left(\lambda^T \bar{h}_N - \psi(\theta, \lambda)\right)$$

is the empirical expectation

$$\bar{h}_N = E_N\left(h(Y)\right) = \frac{1}{N}\sum_{k=1}^{N} h(y_k). \tag{3.98}$$

Together with the number of observations N, the empirical expectation $\bar{h}_N$ is sufficient also for restoration of the likelihood function

$$l_N(\theta) = C' \exp\left(-N \max_\lambda \left(\lambda^T \bar{h}_N - \psi(\theta,\lambda)\right)\right)$$

and the posterior density

$$p_N(\theta) \propto p_0(\theta) \exp\left(-N \max_\lambda \left(\lambda^T \bar{h}_N - \psi(\theta,\lambda)\right)\right).$$

The above formulae follow directly from (2.23) and (2.24).

Note that the sample average $\bar{h}_N$ can easily be computed recursively

$$\begin{aligned}
\bar{h}_N &= \frac{1}{N} \sum_{k=1}^{N} h(y_k) \\
&= \frac{N-1}{N} \bar{h}_{N-1} + \frac{1}{N} h(y_N)
\end{aligned}$$

which makes recursive implementation of estimation immediately possible.

Sufficient Statistic. Consider a family of distributions $\mathcal{S} = \{s_\theta(y) : \theta \in \mathcal{T}\}$. A function

$$T_N : \mathcal{Y}^N \to \mathbb{R}^n$$

(such that $T_N(Y^N)$ is a random variable again, i.e., Borel measurable) is called a *statistic* if it does not depend on θ. A statistic T_N is said to be *sufficient* with respect to the family $\mathcal{S}$ if the sample Y^N and the parameter Θ (taken as a random variable) are conditionally independent given the statistic value $T_N(Y^N) = T_N(y^N)$

$$Y^N \perp \Theta \mid T_N(Y^N) \tag{3.99}$$

for every prior density $p(\theta)$.

In terms of density functions, the condition (3.99) implies

$$q(y^N | T_N(y^N), \theta) = q(y^N | T_N(y^N)) \tag{3.100}$$

and

$$p(\theta | y^N) = p(\theta | T_N(y^N)) \tag{3.101}$$

where in the latter we used the obvious identity $p(\theta|y^N, T_N(y^N)) = p(\theta|y^N)$.

The former coincides with the classical definition of sufficiency. It says essentially that we can construct a sample equivalent to the original one without knowing the actual value of θ. In other words, the value of the statistic $T_N(Y^N)$ yields sufficient information on the distribution of Y^N. The latter is known as a Bayesian definition of sufficiency. According to this, the conditional distribution of Θ depends on the sample Y^N only through the value of the statistic $T_N(Y^N)$.

Note that the two (frequentist and Bayesian) definitions of sufficiency coincide under regularity assumptions made in Sect. 3.3.

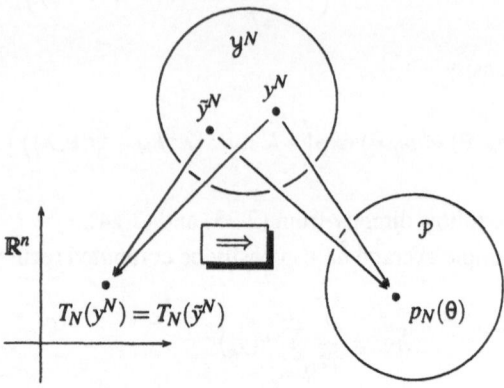

Fig. 3.12. The Bayesian view of sufficient statistic for independent observations. The statistic T_N is sufficient for the parametric family S if for every y^N and $\tilde{y}^N$ such that $T_N(y^N) = T_N(\tilde{y}^N)$ the corresponding posterior densities coincide $p(\theta|y^N) = p(\theta|\tilde{y}^N)$.

Verification of Sufficiency. Let A be the set of all $\tilde{y}^N \in \mathcal{Y}^N$ equivalent with the observed sample y^N in the sense that both give the same value of the statistic

$$T_N(\tilde{y}^N) = T_N(y^N)$$

where the statistic is considered to be a sample average of a single-data vector statistic $h: \mathcal{Y} \to \mathbb{R}^n$

$$T_N(y^N) = \frac{1}{N} \sum_{k=1}^{N} h(y_k) = \bar{h}_N.$$

Suppose that the theoretical densities $s_\theta(y)$ for all $\theta \in \mathcal{T}$ belong to an exponential family (3.40) with the canonical statistic $h(y)$ so that given a particular value $\bar{h}_N$ of the statistic, the h-projection $s_{\theta,\hat{\lambda}}(y)$ of $r_N(y)$ onto the family is common for all $\theta \in \mathcal{T}$. Then by elementary rules of probability theory and the Pythagorean relationship (3.31) we obtain

$$q(y^N, \theta | T_N(y^N))$$

$$= \frac{q_\theta^N(y^N) p_0(\theta)}{\int_A \int_{\mathcal{T}} q_\theta^N(y^N) p_0(\theta) \, d\theta \, dy^N}$$

$$= \frac{\exp(-NK(r_N:s_\theta)) p_0(\theta)}{\int_A \int_{\mathcal{T}} \exp(-NK(r_N:s_\theta)) p_0(\theta) \, d\theta \, dy^N}$$

$$= \frac{\exp(-NK(r_N:s_{\theta,\hat{\lambda}})) \exp(-ND(s_{\theta,\hat{\lambda}} \| s_\theta)) p_0(\theta)}{\int_A \int_{\mathcal{T}} \exp(-NK(r_N:s_{\theta,\hat{\lambda}})) \exp(-ND(s_{\theta,\hat{\lambda}} \| s_\theta)) p_0(\theta) \, d\theta \, dy^N}$$

$$= \frac{\exp(-NK(r_N:s_{\theta,\hat{\lambda}}))}{\int_A \exp(-NK(r_N:s_{\theta,\hat{\lambda}})) \, dy^N} \frac{\exp(-ND(s_{\theta,\hat{\lambda}} \| s_\theta)) p_0(\theta)}{\int_{\mathcal{T}} \exp(-ND(s_{\theta,\hat{\lambda}} \| s_\theta)) p_0(\theta) \, d\theta}.$$

The assumption (3.40) implies that $K(r_N:s_{\theta,\hat{\lambda}})$ is independent of θ and $D(s_{\theta,\hat{\lambda}} \| s_\theta)$ is independent, given the statistic value $\bar{h}_N$, of particular y^N. As a result, we have

$$q(y^N, \theta \,|\, T_N(y^N)) = q(y^N \,|\, T_N(y^N)) \, p(\theta \,|\, T_N(y^N))$$

with

$$q(y^N \,|\, T_N(y^N)) = \frac{\exp(-NK(r_N:s_{\theta,\hat{\lambda}}))}{\int_A \exp(-NK(r_N:s_{\theta,\hat{\lambda}})) \, dy^N},$$

$$p(\theta \,|\, T_N(y^N)) = \frac{\exp(-ND(s_{\theta,\hat{\lambda}} \| s_\theta)) \, p_0(\theta)}{\int_{\mathcal{T}} \exp(-ND(s_{\theta,\hat{\lambda}} \| s_\theta)) \, p_0(\theta) \, d\theta}.$$

Hence, the statistic T_N is sufficient for the model family $\mathcal{S} = \{s_\theta(y) : \theta \in \mathcal{T}\}$.

Note the role of the Pythagorean relationship in the above. We used the fact that under the assumption that $\mathcal{S}$ is imbedded in an exponential family with the canonical statistic $h(y)$, the inaccuracy can be decomposed into a sum of two terms one of which depends on data only while the other, which depends on the parameter, is affected by data merely through the statistic value

$$\underbrace{\exp(-NK(r_N:s_\theta))}_{\phi(y^N, \theta)} = \underbrace{\exp(-NK(r_N:s_{\theta,\hat{\lambda}}))}_{\phi_1(y^N)} \cdot \underbrace{\exp(-ND(s_{\theta,\hat{\lambda}} \| s_\theta))}_{\phi_2(\bar{h}_N, \theta)}.$$

Clearly, the above is nothing but another expression of the factorization criterion of sufficiency well known in statistical theory.

The above also shows that the statistic value $\bar{h}_N$, the set $\mathcal{R}_N$ of densities $r(y)$ compatible with the statistic value and the projection $s_{\theta,\hat{\lambda}}(y)$ bring essentially the same information and can be regarded as equivalent descriptions of compressed data.

Minimal Sufficient Statistic. A statistic $T_N(Y^N)$ is called *minimal sufficient* if it is a function of any other sufficient statistic $T_N'(Y^N)$. Under our regularity assumptions, the minimal sufficient statistic can easily be found in the following way, which essentially dates back to Dynkin (1951), cf. also Zacks (1971).

Suppose $\mathcal{H}$ is the minimal vector space of functions defined on $\mathcal{Y}$ that contains constants and functions

$$\log s_\theta(y) - \log s_{\theta_0}(y)$$

for all $\theta \in \mathcal{T}$ where θ_0 is an arbitrary fixed point in $\mathcal{T}$.

Let $\mathcal{H}$ be spanned by the following set of linearly independent functions

$$h_0(y) \equiv 1, h_1(y), \dots, h_n(y).$$

Then for every θ there exists $\lambda_0 \in \mathbb{R}$ and $\lambda \in \mathbb{R}^n$ such that

$$\log s_\theta(y) = \log s_{\theta_0}(y) + \lambda_0(\theta) + \sum_{i=1}^{n} \lambda_i(\theta) h_i(y).$$

This implies

$$\int s_\theta(y) \, dy = \exp(\lambda_0(\theta)) \int s_{\theta_0}(y) \exp\left(\sum_{i=1}^{n} \lambda_i(\theta) h_i(y)\right) dy.$$

As $\int s_\theta(y) \, dy = 1$, logarithm of the normalizing divisor is

$$\psi(\lambda(\theta)) = \log \int s_{\theta_0}(y) \exp\left(\sum_{i=1}^{n} \lambda_i(\theta) h_i(y)\right) dy$$

$$= -\lambda_0(\theta)$$

and we can write

$$s_\theta(y) = s_{\theta_0}(y) \exp\left(\sum_{i=1}^{n} \lambda_i(\theta) h_i(y) - \psi(\lambda(\theta))\right). \tag{3.102}$$

It is easy to see that $E_N(h(Y))$ forms a sufficient statistic because

$$\prod_{k=1}^{N} s_\theta(y_k) = \prod_{k=1}^{N} s_{\theta_0}(y_k) \exp\left(\sum_{i=1}^{n} \lambda_i(\theta) h_i(y_k) - \psi(\lambda(\theta))\right)$$

$$= \left(\prod_{k=1}^{N} s_{\theta_0}(y_k)\right) \exp\left(N\left(\sum_{i=1}^{n} \lambda_i(\theta) \frac{1}{N} \sum_{k=1}^{N} h_i(y_k) - \psi(\lambda(\theta))\right)\right)$$

$$= \left(\prod_{k=1}^{N} s_{\theta_0}(y_k)\right) \exp\left(N\left(\sum_{i=1}^{n} \lambda_i(\theta) \bar{h}_{i,N} - \psi(\lambda(\theta))\right)\right).$$

The minimality of the statistic follows, under our regularity assumptions, from the linear independence of $h_1(y), \ldots, h_n(y)$. The dimension of the minimal sufficient statistic is thus n.

Compare (3.102) with (3.40). Clearly, the geometric requirement to make $K(r_N : s_{\theta, \hat{\lambda}})$ independent of θ led us to the same conclusion as the statistical requirement to compress data without loss of any information necessary for restoration of the likelihood function and posterior density.

Example 3.6 (*Normal distribution*) Given the normal sampling density

$$s_\theta(y) = \frac{1}{\sqrt{2\pi}} \exp\left(-\frac{1}{2}(y - \theta)^2\right)$$

we have

$$\log s_\theta(y) - \log s_{\theta_0}(y)$$

$$= -\frac{1}{2}\log 2\pi - \frac{1}{2}(y - \theta)^2 + \frac{1}{2}\log 2\pi - \frac{1}{2}(y - \theta_0)^2$$

$$= (\theta - \theta_0)y - \frac{1}{2}(\theta^2 - \theta_0^2).$$

Clearly, for estimation of θ it is sufficient to compute the empirical expectation of

$$h_1(Y) = Y$$

which is just a sample average or arithmetic mean of observed data.   □

Possible Constructions of Canonical Statistic. The above construction of a canonical statistic $h(y)$ of the exponential family enveloping S requires to construct the linear space $\mathcal{H}$ first and then to look for its basis. This is usually cumbersome for practical purposes.

We show here more direct ways of constructing the statistics $h_1(y), \ldots, h_n(y)$.

Differencing. Pick up $n+1$ points $\theta_1^*, \ldots, \theta_{n+1}^*$ in the parameter space $\mathcal{T}$ and set

$$h_j^*(y) = \log s_{\theta_{j+1}^*}(y) - \log s_{\theta_j^*}(y).$$ (3.103)

When the functions $h_0(y) \equiv 1, h_1^*(y), \ldots, h_n^*(y)$ are linearly independent, the vector function $h^*(y) = [h_1^*(y), \ldots, h_n^*(y)]^T$ forms a canonical statistic which is a linear transform of the canonical statistic $h(y) = [h_1(y), \ldots, h_n(y)]^T$ in (3.102)

$$
\begin{aligned}
h_j^*(y) &= \log \frac{s_{\theta_0}(y) \exp\left(\sum_{i=1}^n \lambda_i(\theta_{j+1}^*) h_i(y) - \psi(\lambda(\theta_{j+1}^*)) \right)}{s_{\theta_0}(y) \exp\left(\sum_{i=1}^n \lambda_i(\theta_j^*) h_i(y) - \psi(\lambda(\theta_j^*)) \right)} \\
&= \sum_{i=1}^n \left(\lambda_i(\theta_{j+1}^*) - \lambda_i(\theta_j^*) \right) h_i(y) - \psi(\lambda(\theta_{j+1}^*)) + \psi(\lambda(\theta_j^*)).
\end{aligned}
$$

Differentiation. Suppose that $\log s_\theta(y)$ is differentiable at every $\theta \in \mathcal{T}$ and for all $y \in \mathcal{Y}$. Pick up n points $\theta_1^*, \ldots, \theta_n^*$ in the parameter space $\mathcal{T}$ and n column vectors $\omega_1^*, \ldots, \omega_n^*$ from $\mathbb{R}^{\dim \theta}$. Set

$$h_j^*(y) = \omega_j^{*T} \nabla_\theta \log s_{\theta_j^*}(y).$$ (3.104)

When the functions $h_0(y) \equiv 1, h_1^*(y), \ldots, h_n^*(y)$ are linearly independent, the vector function $h^*(y) = [h_1^*(y), \ldots, h_n^*(y)]^T$ forms a canonical statistic which is a linear transform of the canonical statistic $h(y) = [h_1(y), \ldots, h_n(y)]^T$ in (3.102)

$$
\begin{aligned}
h_j^*(y) &= \omega_j^{*T} \nabla_\theta \log \left(s_{\theta_0}(y) \exp\left(\sum_{i=1}^n \lambda_i(\theta_j^*) h_i(y) - \psi(\lambda(\theta_j^*)) \right) \right) \\
&= \sum_{i=1}^n \left(\omega_j^{*T} \nabla_\theta \lambda_i(\theta_j^*) \right) h_i(y) - \left(\omega_j^{*T} \nabla_\theta \psi(\lambda(\theta_j^*)) \right).
\end{aligned}
$$

Weighted Integration. Pick up n weighting functions $w_1^*(\theta), \ldots, w_n^*(\theta)$ such that

$$\int w_j^*(\theta) \, d\theta = 0, \quad j = 1, \ldots, n.$$

Set

$$h_j^*(y) = \int w_j^*(\theta) \log s_\theta(y) \, d\theta.$$ (3.105)

When the functions $h_0(y) \equiv 1$, $h_1^*(y)$, ..., $h_n^*(y)$ are linearly independent, the vector function $h^*(y) = [h_1^*(y), \ldots, h_n^*(y)]^T$ forms a canonical statistic which is a linear transform of the canonical statistic $h(y) = [h_1(y), \ldots, h_n(y)]^T$ in (3.102)

$$
h_j^*(y) = \int w_j^*(\theta) \log\left(s_{\theta_0}(y) \exp\left(\sum_{i=1}^n \lambda_i(\theta) h_i(y) - \psi(\lambda(\theta)) \right) \right) d\theta
$$

$$
= \sum_{i=1}^n \left(\int w_j^*(\theta) \lambda_i(\theta) d\theta \right) h_i(y) - \left(\int w_j^*(\theta) \psi(\lambda(\theta)) d\theta \right).
$$

Example 3.7 (*Normal distribution*) Given the normal sampling density

$$
s_\theta(y) = \frac{1}{\sqrt{2\pi}} \exp\left(-\frac{1}{2}(y-\theta)^2 \right)
$$

the definitions (3.103), (3.104), (3.105) yield

$$
h_j^*(y) = (\theta_{j+1}^* - \theta_j^*)y - \frac{1}{2}(\theta_{j+1}^{*2} - \theta_j^{*2}),
$$
$$
h_j^*(y) = y - \theta_j^*,
$$
$$
h_j^*(y) = \left(\int w_j^*(\theta) \theta \, d\theta \right) y - \frac{1}{2} \left(\int w_j^*(\theta) \theta^2 \, d\theta \right),
$$

respectively. Clearly, in all the three cases the statistics $h_j^*(y)$ are linear transforms of $h_1(y) = y$. □

General Construction of Canonical Statistic. Other constructions might be used as well like higher-order differencing, higher-order differentiation, combinations of differentiation and weighted integration, etc. It is not difficult to see that all these constructions of a canonical statistic $h^*(y) = [h_1^*(y), \ldots, h_n^*(y)]^T$ have something in common. We show a general construction which covers all the above mentioned cases.

Consider a vector space $\mathcal{H}$ that contains functions

$$
\tilde{h}(\theta) = \log s_\theta(y)
$$

for all $\theta \in \mathcal{T}$. Let L_j^*, $j = 1, \ldots, n$ be linear functionals defined on the vector space $\mathcal{H}$. Suppose in addition that the linear functionals are normalized so that

$$
L_j^*(1) = 0
$$

for $j = 1, \ldots, n$. Then define

$$
\boxed{h_j^*(y) = L_j^*(\log s_\theta(y))} \tag{3.106}
$$

for $j = 1, \ldots, n$.

When the functions $h_0(y) \equiv 1$, $h_1^*(y)$, ..., $h_n^*(y)$ obtained in this way are linearly independent, the vector function $h^*(y) = [h_1^*(y), \ldots, h_n^*(y)]^T$ forms a canonical statistic which is a linear transform of the canonical statistic $h(y) = [h_1(y), \ldots, h_n(y)]^T$ in (3.102)

$$h_j^*(y) = L_j^* \left(\log s_{\theta_0}(y) + \sum_{i=1}^{n} \lambda_i(\theta) h_i(y) - \psi(\lambda(\theta)) \right)$$

$$= \sum_{i=1}^{n} L_j^* \left(\lambda_i(\theta) \right) h_i(y) - L_j^* \left(\psi(\lambda(\theta)) \right).$$

Interpretation of Minimal Sufficient Statistic. Taking into account that

$$E_N \left(\log s_\theta(Y) \right) = \frac{1}{N} \log l_N(\theta),$$

we obtain for the empirical expectation of the single-data statistics (3.103), (3.104), (3.105)

$$E_N \left(\log \frac{s_{\theta_{j+1}^*}(Y)}{s_{\theta_j^*}(Y)} \right) = \frac{1}{N} \log \frac{l_N(\theta_{j+1}^*)}{l_N(\theta_j^*)},$$

$$E_N \left(\omega_j^{*T} \nabla_\theta \log s_{\theta_j^*}(Y) \right) = \frac{1}{N} \omega_j^{*T} \nabla_\theta \log l_N(\theta_j^*),$$

$$E_N \left(\int w_j^*(\theta) \log s_\theta(Y) \, d\theta \right) = \frac{1}{N} \int w_j^*(\theta) \log l_N(\theta) \, d\theta,$$

respectively.

In general, using the definition (3.106), we have

$$E_N \left(L_j^* \left(\log s_\theta(Y) \right) \right) = \frac{1}{N} L_j^* \left(\log l_N(\theta) \right).$$

The minimal sufficient statistic thus brings through various linear functionals $L_j^*(\cdot)$ condensed information about the "shape" of the log-likelihood $\log l_N(\theta)$. For instance, using the definition (3.103), $\bar{h}_N$ is basically equivalent (up to an additive constant) to knowledge of the log-likelihood values on a fixed grid of points $\theta_1^*, \ldots, \theta_{n+1}^* \in \mathcal{T}$.

Controlled Dynamic Systems

Formally, the situation is very much like in the preceding case.

Under the condition (3.89), all information which is required from data to reconstruct $K(r_N : s_\theta)$ through (3.90)

$$K(r_N : s_\theta) = C + \max_\lambda \left(\lambda^T \bar{h}_N - \psi(\theta, \lambda) \right)$$

is the empirical expectation

$$\bar{h}_N = E_N \left(h(Y, Z) \right) = \frac{1}{N} \sum_{k=m+1}^{N+m} h(y_k, z_k).$$

Together with the number of observations N, the empirical expectation $\bar{h}_N$ is sufficient also for restoration of the likelihood function

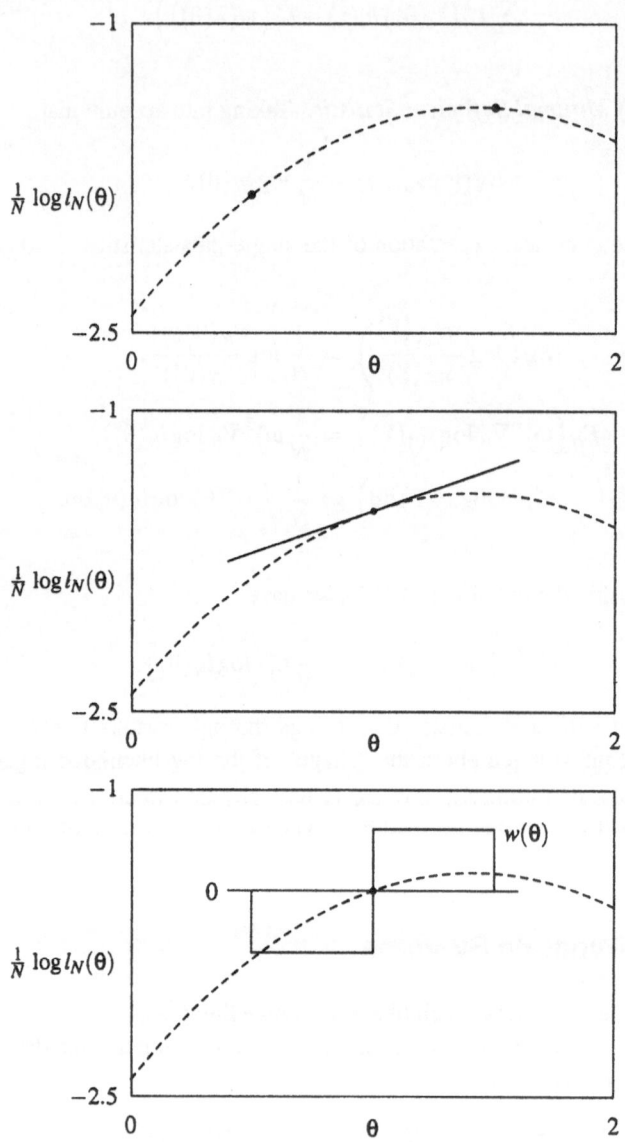

Fig. 3.13. Various ways of constructing a sufficient statistic: values of log-likelihood at prespecified points, derivative of log-likelihood at a particular point, weighted integral of log-likelihood using a zero-integral weighting function. The underlying log-likelihood correspond to $N = 100$ data sampled from the normal distribution $N(1.5, 1)$.

$$l_N(\theta) = C' \exp\left(-N \max_\lambda \left(\lambda^T \bar{h}_N - \psi(\theta,\lambda)\right)\right)$$

and the posterior density

$$p_N(\theta) \propto p_0(\theta) \exp\left(-N \max_\lambda \left(\lambda^T \bar{h}_N - \psi(\theta,\lambda)\right)\right).$$

The formulae follow directly from (2.33) and (2.34).

The sample average $\bar{h}_N$ can easily be computed recursively

$$\begin{aligned}
\bar{h}_N &= \frac{1}{N} \sum_{k=m+1}^{N+m} h(y_k, z_k) \\
&= \frac{N-1}{N} \bar{h}_{N-1} + \frac{1}{N} h(y_{N+m}, z_{N+m}).
\end{aligned}$$

This is all what is needed for recursive implementation of the above scheme.

Sufficient Statistic. Consider a family of distributions $\mathcal{S} = \{s_\theta(y|z) : \theta \in \mathcal{T}\}$. A statistic

$$T_N \colon \mathcal{Y}^{N+m} \times \mathcal{U}^{N+m} \to \mathbb{R}^n$$

is called *sufficient* with respect to $\mathcal{S}$ if the sample Y^{N+m}, U^{N+m} and the parameter Θ (taken as a random variable) are conditionally independent given the value of the statistic $T_N(Y_{m+1}^{N+m}, U_{m+1}^{N+m}) = T_N(y_{m+1}^{N+m}, u_{m+1}^{N+m})$ and the initial values $Y^m = y^m$, $U^m = u^m$

$$Y_{m+1}^{N+m}, U_{m+1}^{N+m} \perp \Theta \mid T_N(Y^{N+m}, U^{N+m}), Y^m, U^m \tag{3.107}$$

for every prior density $p(\theta|y^m, u^m)$.

In terms of density functions, the condition (3.107) implies

$$\begin{aligned}
&q(y_{m+1}^{N+m}, u_{m+1}^{N+m} | T_N(y^{N+m}, u^{N+m}), y^m, u^m, \theta) \\
&\quad = q(y_{m+1}^{N+m}, u_{m+1}^{N+m} | T_N(y^{N+m}, u^{N+m}), y^m, u^m) \tag{3.108}
\end{aligned}$$

and

$$p(\theta|y^{N+m}, u^{N+m}) = p(\theta | T_N(y^{N+m}, u^{N+m}), y^m, u^m) \tag{3.109}$$

where in the latter we used the obvious identity

$$p(\theta|y_{m+1}^{N+m}, u_{m+1}^{N+m}, T_N(y^{N+m}, u^{N+m}), y^m, u^m) = p(\theta|y^{N+m}, u^{N+m}).$$

The former is the classical definition of sufficiency. It says basically that we can produce a sample equivalent to the original one without knowing the actual value of θ. Hence, the value of the statistic $T_N(Y^{N+m}, U^{N+m})$ carries sufficient information (given the initial values $Y^m = y^m$, $U^m = u^m$) about the distribution of $(Y_{m+1}^{N+m}, U_{m+1}^{N+m})$. The latter is known as a Bayesian definition of sufficiency. If T_N is sufficient, the conditional distribution of Θ depends on the sample Y^{N+m}, U^{N+m} only through the value of the statistic $T_N(Y^{N+m}, U^{N+m})$ (given the initial values $Y^m = y^m$, $U^m = u^m$ again).

Note that the two (frequentist and Bayesian) definitions of sufficiency coincide under regularity conditions we assume here.

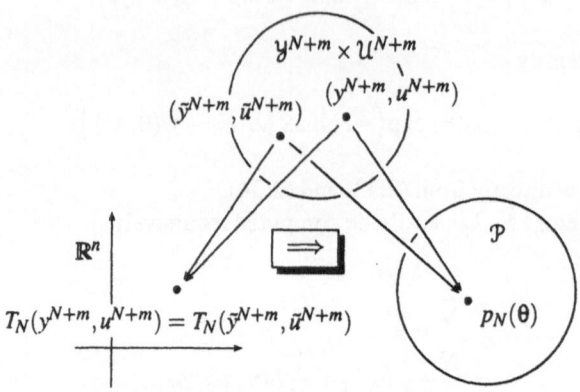

Fig. 3.14. The Bayesian view of sufficient statistic for controlled dynamic systems. The statistic T_N is sufficient for the parametric family S if for every (y^{N+m}, u^{N+m}) and $(\bar{y}^{N+m}, \bar{u}^{N+m})$ such that $T_N(\bar{y}^{N+m}, \bar{u}^{N+m}) = T_N(y^{N+m}, u^{N+m})$ the corresponding posterior densities coincide $p(\theta|\bar{y}^{N+m}, \bar{u}^{N+m}) = p(\theta|y^{N+m}, u^{N+m})$.

Verification of Sufficiency. Let A be the set of all $(\bar{y}_{m+1}^{N+m}, \bar{u}_{m+1}^{N+m}) \in \mathcal{Y}_{m+1}^{N+m} \times \mathcal{U}_{m+1}^{N+m}$ equivalent with the observed sample $(y_{m+1}^{N+m}, u_{m+1}^{N+m})$ in the sense that both yield, given the initial values y^m, u^m, the same value of the statistic

$$T_N(\bar{y}^{N+m}, \bar{u}^{N+m}) = T_N(y^{N+m}, u^{N+m})$$

where the statistic is considered to be a sample average of a single-data vector statistic $h: \mathcal{Y} \times \mathcal{Z} \to \mathbb{R}^n$

$$T_N(y^{N+m}, u^{N+m}) = \frac{1}{N} \sum_{k=m+1}^{N+m} h(y_k, z_k) = \bar{h}_N.$$

Suppose that the theoretical densities $s_\theta(y|z)$ for all $\theta \in \mathcal{T}$ belong, up to a multiplicative constant, to a jointly exponential family (3.89) with the canonical statistic $h(y, z)$ so that given a particular value $\bar{h}_N$ of the statistic, the h-projection $s_{\theta, \hat{\lambda}}(y, z)$ of $r_N(y, z)$ onto the family is common for all $\theta \in \mathcal{T}$. By elementary rules of probability theory and the Pythagorean relationship (3.81) we obtain then

$$q(y_{m+1}^{N+m}, u_{m+1}^{N+m}, \theta | T_N(y_{m+1}^{N+m}, u_{m+1}^{N+m}), y^m, u^m)$$

$$= \frac{q_\theta^N(y_{m+1}^{N+m}, u_{m+1}^{N+m} | y^m, u^m) p_0(\theta)}{\int_A \int_{\mathcal{T}} q_\theta^N(y_{m+1}^{N+m}, u_{m+1}^{N+m} | y^m, u^m) p_0(\theta) \, d\theta \, dy_{m+1}^{N+m} \, du_{m+1}^{N+m}}$$

$$= \frac{\prod_{k=m+1}^{N+m} s_\theta(y_k|z_k) \prod_{k=m+1}^{N+m} \gamma_k(u_k|y^{k-1}, u^{k-1}) p_0(\theta)}{\int_A \int_{\mathcal{T}} \prod_{k=m+1}^{N+m} s_\theta(y_k|z_k) \prod_{k=m+1}^{N+m} \gamma_k(u_k|y^{k-1}, u^{k-1}) p_0(\theta) \, d\theta \, dy_{m+1}^{N+m} \, du_{m+1}^{N+m}}$$

$$= \exp(-NK(r_N : s_{\theta, \hat{\lambda}})) \exp(-ND(s_{\theta, \hat{\lambda}} \| s_\theta)) \prod_{k=m+1}^{N+m} \gamma_k(u_k|y^{k-1}, u^{k-1}) p_0(\theta)$$

$$\cdot \left[\int_A \int_{\mathcal{T}} \exp(-NK(r_N : s_{\theta, \hat{\lambda}})) \exp(-ND(s_{\theta, \hat{\lambda}} \| s_\theta)) \right.$$

$$\cdot \prod_{k=m+1}^{N+m} \gamma_k(u_k|y^{k-1}, u^{k-1}) \, p_0(\theta) \, d\theta \, dy_{m+1}^{N+m} \, du_{m+1}^{N+m} \Big]^{-1}$$

$$= \frac{\exp\big(-NK(r_N{:}s_{\theta,\hat{\lambda}})\big) \prod_{k=m+1}^{N+m} \gamma_k(u_k|y^{k-1}, u^{k-1})}{\int_A \exp\big(-NK(r_N{:}s_{\theta,\hat{\lambda}})\big) \prod_{k=m+1}^{N+m} \gamma_k(u_k|y^{k-1}, u^{k-1}) \, dy_{m+1}^{N+m} \, du_{m+1}^{N+m}}$$

$$\cdot \frac{\exp\big(-ND(s_{\theta,\hat{\lambda}}\|s_\theta)\big) \, p_0(\theta)}{\int_{\mathcal{J}} \exp\big(-ND(s_{\theta,\hat{\lambda}}\|s_\theta)\big) \, p_0(\theta) \, d\theta}.$$

Owing to the assumption (3.89), $K(r_N{:}s_{\theta,\hat{\lambda}})$ is independent of θ and $D(s_{\theta,\hat{\lambda}}\|s_\theta)$ is independent, given the statistic value $\bar{h}_N$, of particular (y^{N+m}, u^{N+m}). As a result, we have

$$q(y_{m+1}^{N+m}, u_{m+1}^{N+m}, \theta \,|\, T_N(y^{N+m}, u^{N+m}), y^m, u^m)$$
$$= q(y_{m+1}^{N+m}, u_{m+1}^{N+m} \,|\, T_N(y^{N+m}, u^{N+m}), y^m, u^m)$$
$$\cdot p(\theta \,|\, T_N(y^{N+m}, u^{N+m}), y^m, u^m)$$

with

$$q(y_{m+1}^{N+m}, u_{m+1}^{N+m} \,|\, T_N(y^{N+m}, u^{N+m}), y^m, u^m)$$
$$= \frac{\exp\big(-NK(r_N{:}s_{\theta,\hat{\lambda}})\big) \prod_{k=m+1}^{N+m} \gamma_k(u_k|y^{k-1}, u^{k-1})}{\int_A \exp\big(-NK(r_N{:}s_{\theta,\hat{\lambda}})\big) \prod_{k=m+1}^{N+m} \gamma_k(u_k|y^{k-1}, u^{k-1}) \, dy_{m+1}^{N+m} \, du_{m+1}^{N+m}},$$

and

$$p(\theta \,|\, T_N(y^{N+m}, u^{N+m}), y^m, u^m) = \frac{\exp\big(-ND(s_{\theta,\hat{\lambda}}\|s_\theta)\big) \, p_0(\theta)}{\int_{\mathcal{J}} \exp\big(-ND(s_{\theta,\hat{\lambda}}\|s_\theta)\big) \, p_0(\theta) \, d\theta}.$$

Hence, the statistic T_N is sufficient for the model family $\mathcal{S} = \{s_\theta(y|z) : \theta \in \mathcal{J}\}$.

Similarly as in the case of independent observations, the Pythagorean relationship enabled us to decompose the inaccuracy $K(r_N{:}s_\theta) = \bar{K}(r_N{:}s_\theta)$ into a sum of two terms one of which depends on data only while the other, dependent on the parameter, is affected by data merely through the statistic value. Hence, we have

$$\overbrace{\phi(y^{N+m}, u^{N+m}, \theta)}$$

$$\overbrace{\exp\big(-NK(r_N{:}s_\theta)\big) \prod_{k=m+1}^{N+m} \gamma_k(u_k|y^{k-1}, u^{k-1})}$$
$$= \underbrace{\exp\big(-NK(r_N{:}s_{\theta,\hat{\lambda}})\big) \prod_{k=m+1}^{N+m} \gamma_k(u_k|y^{k-1}, u^{k-1})}_{\phi_1(y^{N+m}, u^{N+m})} \cdot \underbrace{\exp\big(-ND(s_{\theta,\hat{\lambda}}\|s_\theta)\big)}_{\phi_2(\bar{h}_N, \theta)}.$$

In addition, we assumed that the natural conditions of control are satisfied so that the term $\prod_{k=m+1}^{N+m} \gamma_k(u_k|y^{k-1}, u^{k-1})$ is independent of θ. Altogether, the above can be regarded as an extension of the factorization criterion of sufficiency to the case of controlled dynamic systems.

It also follows from the above that the statistic value $\bar{h}_N$, the set $\mathcal{R}_N$ of densities $r(y,z)$ compatible with the statistic value and the projection $s_{\theta,\hat{\lambda}}(y,z)$ carry the same information and can be taken as equivalent descriptions of compressed data.

Minimal Sufficient Statistic. A statistic $T_N(Y^{N+m}, U^{N+m})$ is *minimal sufficient* if it is a function of any other sufficient statistic $T'_N(Y^{N+m}, U^{N+m})$. Under our regularity assumptions, the minimal sufficient statistic can be constructed as follows.

Suppose $\mathcal{H}$ is the minimal vector space of functions defined on $\mathcal{Y} \times \mathcal{Z}$ that contains constants and functions

$$\log s_\theta(y|z) - \log s_{\theta_0}(y|z)$$

for all $\theta \in \mathcal{T}$ where θ_0 is an arbitrary fixed point in $\mathcal{T}$.

Let $\mathcal{H}$ be spanned by the following set of linearly independent functions

$$h_0(y, z) \equiv 1, h_1(y, z), \ldots, h_n(y, z).$$

Then for every θ there exists $\lambda_0 \in \mathbb{R}$ and $\lambda \in \mathbb{R}^n$ such that

$$\log s_\theta(y|z) = \log s_{\theta_0}(y|z) + \lambda_0(\theta) + \sum_{i=1}^{n} \lambda_i(\theta) h_i(y, z).$$

This expression implies

$$\int s_\theta(y|z)\, dy = \exp(\lambda_0(\theta)) \int s_{\theta_0}(y|z) \exp\left(\sum_{i=1}^{n} \lambda_i(\theta) h_i(y, z)\right) dy.$$

As $\int s_\theta(y|z)\, dy = 1$, logarithm of the normalizing divisor is

$$\begin{aligned} \psi(\lambda(\theta)) &= \log \int s_{\theta_0}(y) \exp\left(\sum_{i=1}^{n} \lambda_i(\theta) h_i(y, z)\right) dy \\ &= -\lambda_0(\theta) \end{aligned}$$

and we can write

$$s_\theta(y|z) = s_{\theta_0}(y|z) \exp\left(\sum_{i=1}^{n} \lambda_i(\theta) h_i(y, z) - \psi(\lambda(\theta))\right). \tag{3.110}$$

Note the factor $\psi(\lambda(\theta))$ is *independent of z*! Therefore, the exponential density with a canonical statistic $h(y, z)$ deserves to be called *jointly* exponential in both Y and Z rather than just conditionally exponential in Y given Z.

It is easy to see that $E_N(h(Y, Z))$ forms a sufficient statistic because

$$\prod_{k=m+1}^{N+m} s_\theta(y_k|z_k)$$

$$= \prod_{k=m+1}^{N+m} s_{\theta_0}(y_k|z_k) \exp\left(\sum_{i=1}^{n} \lambda_i(\theta) h_i(y_k, z_k) - \psi(\lambda(\theta))\right)$$

$$= \left(\prod_{k=m+1}^{N+m} s_{\theta_0}(y_k|z_k)\right) \exp\left(N\left(\sum_{i=1}^{n} \lambda_i(\theta) \frac{1}{N} \sum_{k=m+1}^{N+m} h_i(y_k, z_k) - \psi(\lambda(\theta))\right)\right)$$

$$= \left(\prod_{k=m+1}^{N+m} s_{\theta_0}(y_k|z_k)\right) \exp\left(N\left(\sum_{i=1}^{n} \lambda_i(\theta) \bar{h}_{i,N} - \psi(\lambda(\theta))\right)\right).$$

The minimality of the statistic follows, under our regularity assumptions, from the linear independence of $h_1(y,z),\ldots,h_n(y,z)$. The dimension of the minimal sufficient statistic is thus n.

Compare (3.110) with (3.89). We can see again that the geometric requirement to make $K(r_N:s_{\theta,\hat{\lambda}})$ independent of θ led us to the same result as the statistical requirement to compress data without loss of any information necessary for restoration of the likelihood function and posterior density.

Example 3.8 (*Linear regression with Gaussian noise*) Given the normal conditional sampling density

$$s_\theta(y|z) = \frac{1}{\sqrt{2\pi}}\exp\left(-\frac{1}{2}(y-\theta z)^2\right)$$

with scalar y and z, we have

$$\log s_\theta(y|z) - \log s_{\theta_0}(y|z)$$
$$= -\frac{1}{2}\log 2\pi - \frac{1}{2}(y-\theta z)^2 + \frac{1}{2}\log 2\pi - \frac{1}{2}(y-\theta_0 z)^2$$
$$= (\theta-\theta_0)yz - \frac{1}{2}(\theta^2-\theta_0^2)z^2$$
$$= \lambda_1 yz + \lambda_2 z^2.$$

Hence, for estimation of θ it is sufficient to compute the empirical expectation of

$$h_1(Y,Z) = YZ,$$
$$h_2(Y,Z) = Z^2.$$

$\square$

Possible Constructions of Canonical Statistic. Analogously as in the case of independent observations, there are more direct ways of constructing a canonical statistic $h(y,z)$ of the exponential family enveloping $\mathcal{S}$. We give some typical examples.

Differencing. Pick up $n+1$ points $\theta_1^*, \ldots, \theta_{n+1}^*$ in the parameter space $\mathcal{T}$ and set

$$\boxed{h_j^*(y,z) = \log s_{\theta_{j+1}^*}(y|z) - \log s_{\theta_j^*}(y|z).} \tag{3.111}$$

When the functions $h_0(y,z) \equiv 1$, $h_1^*(y,z)$, $\ldots$, $h_n^*(y,z)$ are linearly independent, the vector function $h^*(y,z) = [h_1^*(y,z),\ldots,h_n^*(y,z)]^T$ forms a canonical statistic which is a linear transform of the canonical statistic $h(y,z) = [h_1(y,z),\ldots,h_n(y,z)]^T$ in (3.110)

$$h_j^*(y,z) = \log\frac{s_{\theta_0}(y|z)\exp\left(\sum_{i=1}^n \lambda_i(\theta_{j+1}^*)h_i(y,z) - \psi(\lambda(\theta_{j+1}^*))\right)}{s_{\theta_0}(y|z)\exp\left(\sum_{i=1}^n \lambda_i(\theta_j^*)h_i(y,z) - \psi(\lambda(\theta_j^*))\right)}$$
$$= \sum_{i=1}^n (\lambda_i(\theta_{j+1}^*) - \lambda_i(\theta_j^*))h_i(y,z) - \psi(\lambda(\theta_{j+1}^*)) + \psi(\lambda(\theta_j^*)).$$

Differentiation. Suppose that $\log s_\theta(y|z)$ is differentiable at every $\theta \in \mathcal{T}$ and for all $(y,z) \in \mathcal{Y} \times \mathcal{Z}$. Pick up n points $\theta_1^*, \ldots, \theta_n^*$ in the parameter space $\mathcal{T}$ and n column vectors $\omega_1^*, \ldots, \omega_n^*$ from $\mathbb{R}^{\dim \theta}$. Set

$$h_j^*(y,z) = \omega_j^{*T} \nabla_\theta \log s_{\theta_j^*}(y|z). \tag{3.112}$$

When the functions $h_0(y,z) \equiv 1, h_1^*(y,z), \ldots, h_n^*(y,z)$ are linearly independent, the vector function $h^*(y,z) = [h_1^*(y,z), \ldots, h_n^*(y,z)]^T$ forms a canonical statistic which is a linear transform of the canonical statistic $h(y,z) = [h_1(y,z), \ldots, h_n(y,z)]^T$ in (3.110)

$$h_j^*(y,z) = \omega_j^{*T} \nabla_\theta \log\left(s_{\theta_0}(y|z) \exp\left(\sum_{i=1}^n \lambda_i(\theta_j^*) h_i(y,z) - \psi(\lambda(\theta_j^*)) \right) \right)$$

$$= \sum_{i=1}^n \left(\omega_j^{*T} \nabla_\theta \lambda_i(\theta_j^*) \right) h_i(y,z) - \left(\omega_j^{*T} \nabla_\theta \psi(\lambda(\theta_j^*)) \right).$$

Weighted Integration. Pick up n weighting functions $w_1^*(\theta), \ldots, w_n^*(\theta)$ such that

$$\int w_j^*(\theta)\, d\theta = 0, \quad j = 1, \ldots, n.$$

Set

$$h_j^*(y,z) = \int w_j^*(\theta) \log s_\theta(y|z)\, d\theta. \tag{3.113}$$

When the functions $h_0(y,z) \equiv 1, h_1^*(y,z), \ldots, h_n^*(y,z)$ are linearly independent, the vector function $h^*(y,z) = [h_1^*(y,z), \ldots, h_n^*(y,z)]^T$ forms a canonical statistic which is a linear transform of the canonical statistic $h(y,z) = [h_1(y,z), \ldots, h_n(y,z)]^T$ in (3.110)

$$h_j^*(y,z) = \int w_j^*(\theta) \log\left(s_{\theta_0}(y|z) \exp\left(\sum_{i=1}^n \lambda_i(\theta) h_i(y,z) - \psi(\lambda(\theta)) \right) \right) d\theta$$

$$= \sum_{i=1}^n \left(\int w_j^*(\theta) \lambda_i(\theta)\, d\theta \right) h_i(y,z) - \left(\int w_j^*(\theta) \psi(\lambda(\theta))\, d\theta \right).$$

Example 3.9 (*Linear regression with Gaussian noise*) Given the normal conditional sampling density

$$s_\theta(y|z) = \frac{1}{\sqrt{2\pi}} \exp\left(-\frac{1}{2}(y - \theta z)^2 \right),$$

the definitions (3.111), (3.112), (3.113) yield

$$h_j^*(y,z) = (\theta_{j+1}^* - \theta_j^*)\, yz - \frac{1}{2}(\theta_{j+1}^{*2} - \theta_j^{*2})\, z^2,$$

$$h_j^*(y,z) = (y - \theta_j^* z)\, z,$$

$$h_j^*(y,z) = \left(\int w_j^*(\theta) \theta\, d\theta \right) yz - \frac{1}{2}\left(\int w_j^*(\theta) \theta^2\, d\theta \right) z^2,$$

respectively. In all cases, the statistics $h_1^*(y,z), h_2^*(y,z)$ are just linear transforms of $h_1(y,z) = yz, h_2(y,z) = z^2$. □

General Construction of Canonical Statistic. A general construction of a canonical statistic $h^*(y,z) = [h_1^*(y,z),\ldots,h_n^*(y,z)]^T$ looks as follows. Consider a vector space $\tilde{\mathcal{H}}$ that contains functions

$$\tilde{h}(\theta) = \log s_\theta(y|z)$$

for all $\theta \in \mathcal{T}$. Let L_j^*, $j = 1,\ldots,n$ be linear functionals defined on the vector space $\tilde{\mathcal{H}}$. Suppose in addition that the linear functionals are normalized so that

$$L_j^*(1) = 0$$

for $j = 1,\ldots,n$. Then define

$$\boxed{h_j^*(y,z) = L_j^*\left(\log s_\theta(y|z)\right)} \qquad (3.114)$$

for $j = 1,\ldots,n$.

When the functions $h_0(y,z) \equiv 1, h_1^*(y,z), \ldots, h_n^*(y,z)$ are linearly independent, the vector function $h^*(y,z) = [h_1^*(y,z),\ldots,h_n^*(y,z)]^T$ forms a canonical statistic which is a linear transform of the canonical statistic $h(y,z) = [h_1(y,z),\ldots,h_n(y,z)]^T$ in (3.110)

$$h_j^*(y,z) = L_j^*\left(\log s_{\theta_0}(y|z) + \sum_{i=1}^n \lambda_i(\theta) h_i(y,z) - \psi(\lambda(\theta))\right)$$

$$= \sum_{i=1}^n L_j^*(\lambda_i(\theta))\, h_i(y,z) - L_j^*\left(\psi(\lambda(\theta))\right).$$

Interpretation of Minimal Sufficient Statistic. Taking into account that

$$E_N\left(\log s_\theta(y|z)\right) = C + \frac{1}{N}\log l_N(\theta)$$

where C is a constant independent of θ, we obtain for the empirical expectation of the single-data statistics (3.111), (3.112), (3.113)

$$E_N\left(\log \frac{s_{\theta_{j+1}^*}(Y|Z)}{s_{\theta_j^*}(Y|Z)}\right) = \frac{1}{N}\log \frac{l_N(\theta_{j+1}^*)}{l_N(\theta_j^*)},$$

$$E_N\left(\omega_j^{*T}\nabla_\theta \log s_{\theta_j^*}(Y|Z)\right) = \frac{1}{N}\omega_j^{*T}\nabla_\theta \log l_N(\theta_j^*),$$

$$E_N\left(\int w_j^*(\theta) \log s_\theta(Y|Z)\, d\theta\right) = \frac{1}{N}\int w_j^*(\theta) \log l_N(\theta)\, d\theta,$$

respectively.

In general, using the definition (3.114), we have

$$E_N\left(L_j^*(\log s_\theta(Y|Z))\right) = \frac{1}{N}L_j^*\left(\log l_N(\theta)\right).$$

Once again, the minimal sufficient statistic is a function of the log-likelihood $\log l_N(\theta)$; through the linear functionals $L_j^*(\cdot)$ we get condensed information about its "shape".

3.6 Historical Notes

Shannon's Entropy. The concept of entropy appeared first in the formulation of the second law of thermodynamics. Later, Boltzmann (1877) in statistical mechanics showed connection between the macroscopic property of entropy and the microscopic state of the system. In the 1930s, Hartley used the logarithm of the alphabet size as a measure of information suitable for communication. The modern definition as we know it today was given in Shannon's (1948) seminal paper that laid foundations of information theory.

Kullback-Leibler Distance. The concept of K-L distance was defined first by Kullback and Leibler (1951) with reference to the previous work in a similar direction by Jeffreys and Shannon. This information measure can be found in the technical literature under various names: Kullback-Leibler information, information divergence, information deviation, directed divergence, discrimination information, Rényi's information gain, expected weight of evidence, entropy distance, relative entropy, cross-entropy.

After Kullback and Leibler (1951), the concept was analysed by many others, including Kullback (1959), Sanov (1957), Csiszár (1967), Čencov (1972), Amari (1985). An axiomatic characterization of Kullback-Leibler distance was given in Kannappan (1972b; 1972a) and Shore and Johnson (1980).

Note that Kullback-Leibler distance can be regarded as a member of a more general class of information distances introduced independently by Csiszár (1967) and Ali and Silvey (1966).

Fisher Information. The concept was introduced by Fisher (1925) as a measure of information about the unknown parameter contained in a single observation. The connection between Fisher information and Kullback-Leibler distance was shown in Kullback (1959).

Riemannian Geometry of Probability Manifolds. Rao (1945) was the first who introduced the Riemannian metric in a parametric family of probability distributions by using the Fisher information matrix. He also calculated the geodesic distances in some statistical models. Čencov (1972) showed that the Riemannian metrics which are derived from the Fisher information matrix are the only metrics which preserve inner products under certain probabilistically important mappings. He also introduced a family of statistically important affine connections in a parametric family of probability distributions. Efron (1975) and Dawid (1975) pointed out the importance of the curvature of a parametric family in the higher-order asymptotic theory of estimation. Using Čencov's ideas, Campbell (1985) analysed connections between information theory and the differential-geometric view of statistics. Amari (1985) introduced the concept of dual affine connections to explain specific, non-Riemannian features of the geometry of parametric families. Burbea and Rao (1982) analysed the relation between the generalized entropy and distance in spaces of probability distributions. Barndorff-Nielsen et al. (1986) provided a survey of known applications of differential-geometric techniques in statistical theory.

Exponential Family. Fisher (1934) indicated that exponential families are the only families of distributions that yield nontrivial sufficiency reductions. This indication was quickly taken up by Darmois (1935), Koopman (1936) and Pitman (1936) who sought its rigorous mathematical formulations. Exponential families appeared independently in problems of statistical physics (Khintchine, 1943; Kubo, 1965). Models of this kind date back to the work of Maxwell, Boltzmann and Gibbs at the end of the last century.

An account of the general theory of exponential families was given in Kullback (1959) and Lehmann (1959). General analytic approaches to the theory of parameter estimation and hypothesis discrimination for these families were proposed in Linnik (1966; 1967). A number of fundamental problems of the theory were also touched upon in Blackwell and Girshick (1954) and Robbins (1956). Čencov (1964; 1966; 1972) dealt mainly with the "geometric" aspects of the theory.

Barndorff-Nielsen (1978) and Brown (1987) wrote comprehensive monographs devoted solely to the exponential families.

Minimum K-L Distance Projections. The Pythagorean theorem that holds for K-L distances between probability distributions was derived in Čencov (1972), Csiszár (1975) and Amari (1985). The Pythagorean-like relationship that links together inaccuracies and K-L distance was shown in Kulhavý (1995a). Jones (1989) showed that K-L distance is the only smooth information distance from the class of f-divergences introduced by Csiszár (1967) and Ali and Silvey (1966) for which the dual minimum K-L distance projections coincide. The dual minimum K-L distance projections were applied to solve various inverse problems— see Christensen (1989) for statistical estimation, Jones and Byrne (1990) for data compression, pattern classification and cluster analysis, Matsuoka and Ulrych (1986) for modelling and identification. Rich bibliography on the applications of minimum K-L distance projections can be found in Shore and Johnson (1980).

Sufficient Statistic. The concept was introduced in Fisher (1922; 1925). A rigorous measure-theoretic definition was given by Halmos and Savage (1949). Bahadur (1954) extended sufficiency to sequential experiments. Lehmann and Scheffé (1950; 1955) introduced the notion of the minimal sufficient statistic. The famous factorization criterion was proved in various degrees of rigour by Fisher (1922), Neyman (1935) and Halmos and Savage (1949). The Bayesian characterization of sufficiency was suggested by Kolmogorov (1942). An information-theoretic characterization of sufficiency was given in Kullback (1959).

From textbooks on mathematical statistics, Lindgren (1976) gives a readable discussion of sufficient statistics, with no measure theory. Lehmann (1959) and Zacks (1971) both present measure-theoretic treatments of properties of sufficient statistics.

Recursively Computable Statistic. The importance of recursive computability was recognized in the system identification literature quite early. It was studied mainly in connection with the concept of sufficiency, see Bohlin (1970), Davis and Varaiya (1972), Willems (1980). Related concepts of transitive and algebraically transitive statistics can be found in the statistical literature in Bahadur (1954) and Lauritzen (1988).

4. Approximate Estimation with Compressed Data

The picture given in the preceding chapter sets up an ideal which is rarely met when dealing with real-life models. One notable exception is the class of linear normal regression-type models. For models with non-linear dynamics and non-Gaussian stochastics, there is little hope that a sufficient statistic exists of dimension we can afford. A more realistic framework is then needed that would make approximate estimation possible under given constraints. The present section demonstrates that the Pythagorean view of estimation provides us with such a framework.

4.1 Further Compressing Data: Euclidean Case

Given a sequence of data (row vector)

$$y_{m+1}^{N+m} = (y_{m+1}, \ldots, y_{N+m}),$$

we consider the problem of computing, approximately at least, the Euclidean distance squared between the true and model-based sequences of data

$$\left\| y_{m+1}^{N+m} - \hat{y}^N(\theta) \right\|^2$$

with as little information as possible.

We have seen in Sect. 3.1 that if all model points $\hat{y}^N(\theta)$ happen to lie within an n-dimensional hyperplane $\mathcal{H}$ in $\mathbb{R}^N$, then it is sufficient to know the orthogonal projection of y_{m+1}^{N+m} onto $\mathcal{H}$ in order to reconstruct, *with precision up to an additive constant*, the "error function" $\left\| y_{m+1}^{N+m} - \hat{y}^N(\theta) \right\|^2$.

The question not touched yet is what to do when the dimension of $\mathcal{H}$ is too large or even growing with N. What seems to be a natural approximation of the ideal no-loss solution is to consider projections of y_{m+1}^{N+m} onto a hyperplane that "almost" captures all model points $\hat{y}^N(\theta)$.

More specifically, we define for every $\theta \in \mathcal{T}$ a hyperplane in $\mathbb{R}^N$ composed of points

$$\hat{y}^N(\theta, \lambda) = \hat{y}^N(\theta) + \lambda^T H$$

where λ is a column vector of dimension n and H is a matrix of type (n, N) whose rows are formed by certain "basis" vectors h_i, $i = 1, \ldots, n$ of dimension N

$$H = \begin{bmatrix} h_1 \\ h_2 \\ \vdots \\ h_n \end{bmatrix}.$$

The choice of the "basis" vectors h_i, $i = 1, \ldots, n$ is directed by the aim to make $\hat{y}^N(\theta, \lambda)$ "almost" independent of θ. In general, H may depend on the observed data y^{N+m-1}. For the sake of simplicity, H is supposed to be known and fixed here, i.e., not exposed to data compression.

Orthogonal Projection. The least-squares matching of the sequence of observed data y_{m+1}^{N+m} with $\hat{y}^N(\theta, \lambda)$ is defined by minimizing the Euclidean distance squared between both the vectors

$$\min_\lambda \left\| y_{m+1}^{N+m} - \hat{y}^N(\theta, \lambda) \right\|^2.$$

A necessary condition for the optimum $\hat{\lambda}_N$ follows by differentiating the Euclidean distance squared with respect to λ

$$\nabla_\lambda \left\| y_{m+1}^{N+m} - \hat{y}^N(\theta) - \hat{\lambda}_N^T H \right\|^2 = -2 \left(y_{m+1}^{N+m} - \hat{y}^N(\theta) - \hat{\lambda}_N^T H \right) H^T = 0.$$

This implies the condition

$$\left(y_{m+1}^{N+m} - \hat{y}^N(\theta) - \hat{\lambda}_N^T H \right) H^T = 0 \tag{4.1}$$

which can be read "geometrically" so that the vector $y_{m+1}^{N+m} - \hat{y}^N(\theta) - \hat{\lambda}_N^T H$ is orthogonal to all rows in H. In other words, $\hat{y}^N(\theta) + \hat{\lambda}_N^T H$ is an orthogonal projection of y_{m+1}^{N+m} onto the hyperplane $\{ \hat{y}^N(\theta, \lambda) : \lambda \in \mathbb{R}^n \}$. One can easily compute from

$$\left(y_{m+1}^{N+m} - \hat{y}^N(\theta) \right) H^T = \hat{\lambda}_N^T H H^T$$

the solution

$$\hat{\lambda}_N = \left(H H^T \right)^{-1} H \left(y_{m+1}^{N+m} - \hat{y}^N(\theta) \right)^T. \tag{4.2}$$

Pythagorean Relationship. The orthogonal projection $\hat{y}^N(\theta) + \hat{\lambda}_N^T H$ clearly satisfies the Pythagorean relationship. Rewriting the Euclidean distance squared as follows

$$\begin{aligned} \left\| y_{m+1}^{N+m} - \hat{y}^N(\theta) \right\|^2 &= \left\| y_{m+1}^{N+m} - \hat{y}^N(\theta) - \hat{\lambda}_N^T H + \hat{\lambda}_N^T H \right\|^2 \\ &= \left\| y_{m+1}^{N+m} - \hat{y}^N(\theta) - \hat{\lambda}_N^T H \right\|^2 + \left\| \hat{\lambda}_N^T H \right\|^2 \\ &\quad + 2 \left(y_{m+1}^{N+m} - \hat{y}^N(\theta) - \hat{\lambda}_N^T H \right) \left(\hat{\lambda}_N^T H \right)^T \end{aligned}$$

and taking into account (4.1), we obtain

$$\left\| y_{m+1}^{N+m} - \hat{y}^N(\theta) \right\|^2 = \left\| y_{m+1}^{N+m} - \hat{y}^N(\theta) - \hat{\lambda}_N^T H \right\|^2 + \left\| \hat{\lambda}_N^T H \right\|^2. \tag{4.3}$$

The identity says that the Euclidean distance squared between y_{m+1}^{N+m} and the point $\hat{y}^N(\theta)$ is equal to a sum of the Euclidean distance squared between y_{m+1}^{N+m} and its orthogonal projection $\hat{y}^N(\theta) + \hat{\lambda}_N^T H$ and the Euclidean distance squared between $\hat{y}^N(\theta) + \hat{\lambda}_N^T H$ and $\hat{y}^N(\theta)$.

Approximate Estimation. The difference between the present solution and the case considered in Sect. 3.1 lies in the fact that the Pythagorean relationship (4.3) is satisfied now at *different* points $\hat{\lambda}_N$ for *different* values of θ. Consequently, the Euclidean distance squared

$$\left\| y_{m+1}^{N+m} - \hat{y}^N(\theta) - \hat{\lambda}_N^T H \right\|^2$$

is *not* independent of θ any more. Yet, as shown below, it makes a good sense to use approximation

$$\left\| y_{m+1}^{N+m} - \hat{y}^N(\theta) \right\|^2 \approx C + \left\| \hat{y}^N(\theta, \hat{\lambda}_N) - \hat{y}^N(\theta) \right\|^2 = C + \left\| \hat{\lambda}_N^T H \right\|^2 \qquad (4.4)$$

where $C > 0$ is a constant independent of θ. Substituting for $\hat{\lambda}_N$ from (4.2), we obtain

$$\left\| y_{m+1}^{N+m} - \hat{y}^N(\theta) \right\|^2 \approx C + \left(y_{m+1}^{N+m} - \hat{y}^N(\theta) \right) H^T (H H^T)^{-1} H \left(y_{m+1}^{N+m} - \hat{y}^N(\theta) \right)^T.$$

Note that all we need to calculate the right-hand side is the square matrix $H H^T$ of order n, the vector $y_{m+1}^{N+m} H^T$ of dimension n and the vector $\hat{y}^N(\theta) H^T$ of dimension n (for all $\theta \in \mathcal{T}$ which may still be a problem).

The approximation (4.4) is supported by the following inequality that follows directly from the Pythagorean relationship (4.3)

$$0 \le \frac{1}{N} \left\| \hat{\lambda}_N^T H \right\|^2 = \frac{1}{N} \left\| y_{m+1}^{N+m} - \hat{y}^N(\theta) \right\|^2 - \frac{1}{N} \left\| y_{m+1}^{N+m} - \hat{y}^N(\theta, \hat{\lambda}_N) \right\|^2.$$

Suppose that the normalized Euclidean distance squared $\frac{1}{N} \left\| y_{m+1}^{N+m} - \hat{y}^N(\theta) \right\|^2$ attains asymptotically at $\theta = \theta_0$ its absolute minimum—such that cannot be improved by taking the extended model $\hat{y}(\theta, \lambda)$. Then the right-hand side of the above inequality approaches necessarily zero at $\theta = \theta_0$ and, consequently, the approximate estimator is asymptotically *consistent* with the ideal estimator.

In fact, we can say more than that. Let Ψ_N denote the set of all data sequences $\tilde{y}_{m+1}^{N+m}$ the orthogonal projection of which onto $\{\hat{y}^N(\theta, \lambda) : \lambda \in \mathbb{R}^n\}$ coincides with the projection of the true data y_{m+1}^{N+m}, i.e., it holds

$$\tilde{y}_{m+1}^{N+m} H^T = y_{m+1}^{N+m} H^T.$$

Then owing to the Pythagorean relationship (4.3) we have

$$\left\| \hat{\lambda}_N^T H \right\|^2 = \min_{\tilde{y}_{m+1}^{N+m} \in \Psi_N} \left\| \tilde{y}_{m+1}^{N+m} - \hat{y}^N(\theta) \right\|^2$$

$$\le \left\| \tilde{y}_{m+1}^{N+m} - \hat{y}^N(\theta) \right\|^2 \quad \text{for every} \quad \tilde{y}_{m+1}^{N+m} \in \Psi_N.$$

The above inequality provides another justification for the approximation (4.4). Given insufficient information about the observed data, the approximation suggests to take the *minimum distance* between the model-based sequence $\hat{y}^N(\theta)$ and the set Ψ_N of all sequences $\tilde{y}^{N+m}_{m+1}$ which cannot be distinguished from y^{N+m}_{m+1} (cf. Fig. 4.1). The minimum distance represents the least favourable measure of discrimination between data and model.

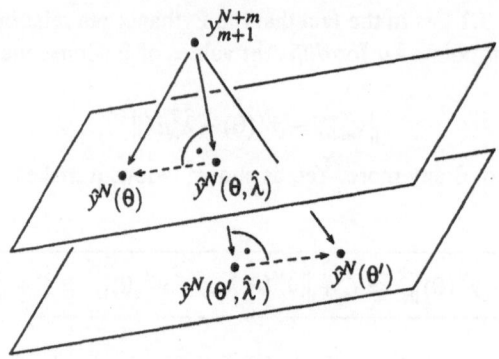

Fig. 4.1. Measuring Euclidean distance between given and model-based sequences of data.

4.2 Idea of Approximate Estimation

The idea of minimum distance approximation naturally extends to probability-based estimation.

Independent Observations

Estimation Problem. Suppose that a certain statistic of Y is chosen

$$h: \mathcal{Y} \to \mathbb{R}^n.$$

Let the only information available about the empirical density $r_N(y)$ be the empirical expectation of $h(Y)$

$$\bar{h}_N \overset{\triangle}{=} \int r_N(y)\, h(y)\, dy = \frac{1}{N} \sum_{k=1}^{N} h(y_k).$$

The empirical density $r_N(y)$ is thus known to lie within the set

$$\mathcal{R}_N \overset{\triangle}{=} \left\{ r(y): \int r(y)\, h(y)\, dy = \bar{h}_N, \int r(y)\, dy = 1, \ r(y) \geq 0 \right\}.$$

The problem is to compute the inaccuracy $K(r_N : s_\theta)$ as a function of θ given the above partial information about $r_N(y)$.

Approximation of Inaccuracy. We start from the Pythagorean relationship (3.31)

$$K(r_N:s_\theta) = K(r_N:s_{\theta,\hat{\lambda}}) + D(s_{\theta,\hat{\lambda}}\|s_\theta).$$

The approximation we propose is to set

$$K(r_N:s_{\theta,\hat{\lambda}}) \approx C \tag{4.5}$$

where C is a constant independent of θ. This makes a good sense; the inaccuracy $K(r_N:s_{\theta,\hat{\lambda}})$ can be made almost independent of θ through a proper choice of $h(y)$ which is discussed in detail in Sect. 4.5.

Next by (3.35) and (3.38), we have

$$D(s_{\theta,\hat{\lambda}}\|s_\theta) = \min_{r \in \mathcal{R}_N} D(r\|s_\theta)$$

$$= \max_{\lambda} \left(\lambda^T \bar{h}_N - \psi(\theta,\lambda) \right).$$

Introducing the notation

$$D(\mathcal{R}_N\|s_\theta) \stackrel{\Delta}{=} \min_{r \in \mathcal{R}_N} D(r\|s_\theta) \tag{4.6}$$

and substituting for $\psi(\theta,\lambda)$ from (3.26), we obtain

$$\boxed{D(\mathcal{R}_N\|s_\theta) = \max_{\lambda} \left(\lambda^T \bar{h}_N - \log \int s_\theta(y) \exp(\lambda^T h(y)) \right).} \tag{4.7}$$

Hence, under (4.5), we have

$$\boxed{K(r_N:s_\theta) \approx C + D(\mathcal{R}_N\|s_\theta).} \tag{4.8}$$

Approximation of Likelihood. Substituting from (4.8) for $K(r_N:s_\theta)$ in (2.23), we obtain the approximate likelihood function

$$\boxed{\hat{l}_N(\theta) = C \exp(-ND(\mathcal{R}_N\|s_\theta)).} \tag{4.9}$$

Similarly, substituting from (4.8) for $K(r_N:s_\theta)$ in (2.24), we obtain the approximate posterior density

$$\boxed{\hat{p}_N(\theta) \propto p_0(\theta) \exp(-ND(\mathcal{R}_N\|s_\theta)).} \tag{4.10}$$

Example 4.1 *(Cauchy distribution)*   We considered a sequence of observations $Y_1,\ldots,Y_N$ modelled through

$$Y_k = \theta + E_k$$

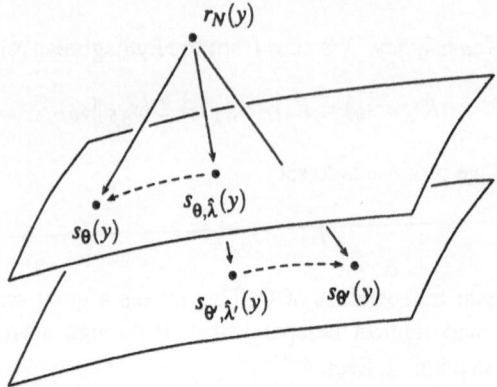

Fig. 4.2. The idea of approximate estimation for independent observations. Instead of computing the inaccuracy $K(r_N : s_\theta)$ directly, the minimum K-L distance $D(\mathcal{R}_N \| s_\theta) = D(s_{\theta,\hat\lambda} \| s_\theta)$ is computed for every $\theta \in \mathcal{T}$.

where $E_1, \ldots, E_N$ was a sequence of independent, Cauchy-distributed random variables with a common density

$$n(e) = \frac{1}{\pi} \frac{1}{1 + e^2}.$$

The theoretical density of Y_k was thus

$$s_\theta(y) = \frac{1}{\pi} \frac{1}{1 + (y - \theta)^2}.$$

The problem solved was to estimate the regression coefficient θ given a particular sequence of 100 data $(y_1, \ldots, y_{100})$ shown in Fig. 4.3.

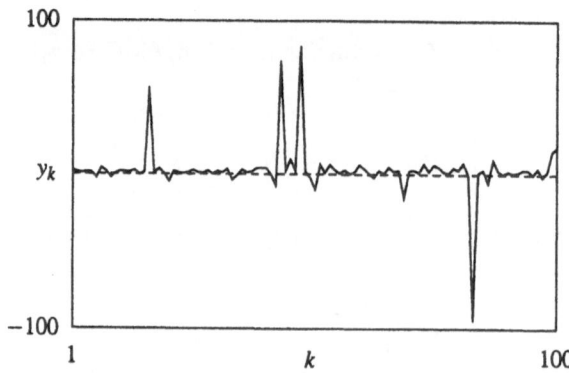

Fig. 4.3. Cauchyian noise: a sequence of 100 samples of $Y_k = 1 + E_k$ with Cauchy-distributed noise $E_k \sim C(0,1)$.

We used a vector statistic $h(y)$ of dimension $n = 3$ composed of score functions—the first-order derivatives of the log-density $\log s_\theta(y)$ with respect to θ

$$h_i(y) = \frac{2(y - \theta_i^*)}{1 + (y - \theta_i^*)^2} \tag{4.11}$$

at the points $\theta_i^* = -2, 0, +2$. The functions h_1, h_2, h_3 are plotted in Fig. 4.4.

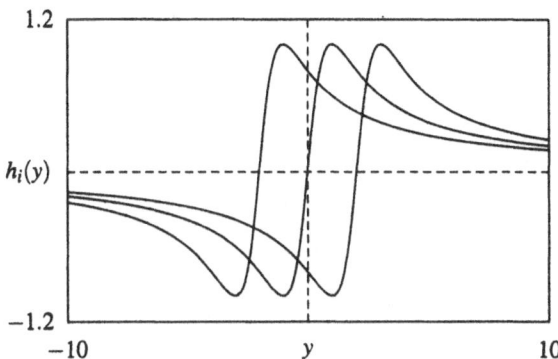

Fig. 4.4. Cauchyian noise: a set of single-data statistics $h_i(y)$ used for data compression.

Given the statistic $h(y)$ and the observed data $(y_1, \ldots, y_{100})$, we computed the sample average

$$\bar{h}_N = \frac{1}{N} \sum_{k=1}^{N} h(y_k)$$

for $N = 100$.

Given the value $\bar{h}_N$, we solved the optimization problem (4.7) for a set of different values of the unknown parameter θ, namely 121 values equidistantly located within the interval $[-6, 6]$. The resulting K-L distance $D(\mathcal{R}_N \| s_\theta)$ is plotted using a solid line in the upper plot in Fig. 4.5.

To illustrate that the K-L distance $D(r \| s_\theta)$ for every $r(y) \in \mathcal{R}_N$ is bounded from below by $D(\mathcal{R}_N \| s_\theta)$, we calculated $D(r_i \| s_\theta)$ for six such densities $r_i(y)$, constructed as minimum K-L distance projections of the theoretical densities $s_{\theta_i^*}(y)$ for $\theta_i^* = -6, -3.6, -1.2, +1.2, +3.6, +6$ onto $\mathcal{R}_N$.

The lower plot in Fig. 4.5 shows the normalized likelihoods

$$\hat{l}_N(\theta) = \exp\big(-N D(\mathcal{R}_N \| s_\theta)\big)$$
$$l_N(\theta; r_i) = \exp\big(-N D(r_i \| s_\theta)\big)$$

for $N = 100$.

Fig. 4.6 gives a more detailed picture of approximate estimation for one particular value of θ. $\qquad\square$

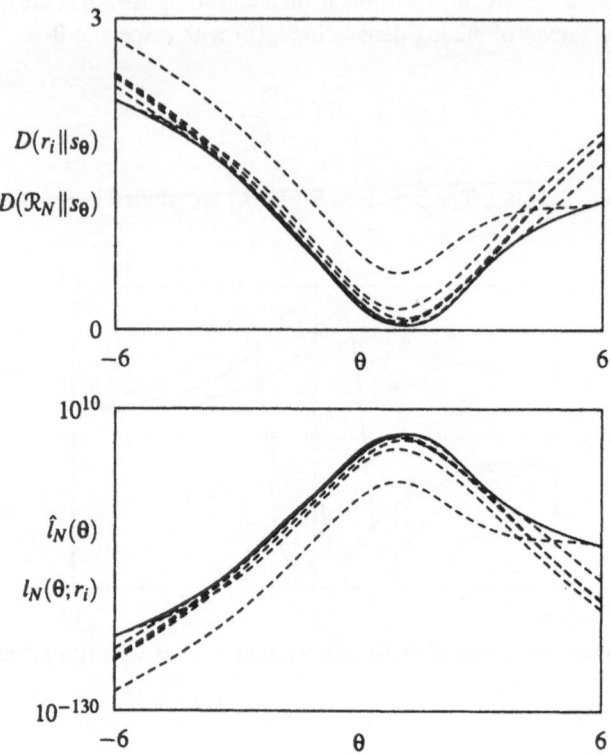

Fig. 4.5. Cauchyian noise: results of approximate estimation for $N = 100$. The upper plot compares K-L distances, the lower plot shows the normalized likelihoods.

Controlled Dynamic Systems

Estimation Problem. Suppose that a certain statistic of (Y, Z) is chosen

$$h: \mathcal{Y} \times \mathcal{Z} \to \mathbb{R}^n.$$

Let the only information available about the empirical density $r_N(y, z)$ be the empirical expectation of $h(Y, Z)$

$$\bar{h}_N \triangleq \iint r_N(y, z) h(y, z) \, dy \, dz = \frac{1}{N} \sum_{k=m+1}^{N+m} h(y_k, z_k).$$

The empirical density $r_N(y, z)$ is thus known to lie within the set

$$\mathcal{R}_N \triangleq \left\{ r(y, z) : \iint r(y, z) h(y, z) \, dy \, dz = \bar{h}_N, \iint r(y, z) \, dy \, dz = 1, \, r(y) \geq 0 \right\}.$$

The problem is to compute the conditional inaccuracy $\bar{K}(r_N : s_\theta)$ as a function of θ given the above partial information about $r_N(y, z)$.

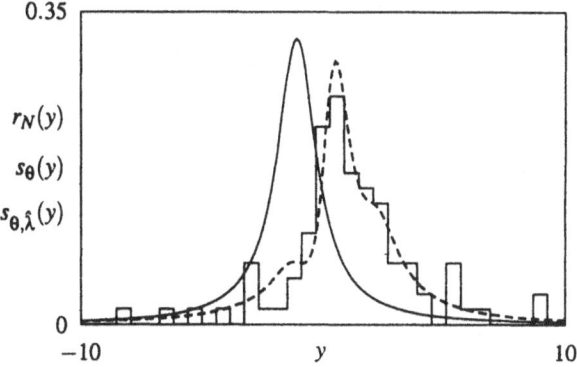

Fig. 4.6. Cauchyian noise: illustration of the principle of approximation for $\theta = -1$. Given the theoretical density $s_\theta(y)$ (solid line) and the statistic value $\bar{h}_N$, the minimum K-L distance approximation $s_{\theta,\hat{\lambda}}(y)$ (dashed line) of the empirical density $r_N(y)$ (shown by histogram) is constructed. The K-L distance $D(s_{\theta,\hat{\lambda}} \| s_\theta)$ is then used to approximate the unmeasurable inaccuracy $K(r_N : s_\theta)$.

Approximation of Inaccuracy. The idea is the same as for independent observations. We start from the Pythagorean relationship (3.81)

$$K(r_N : s_\theta) = K(r_N : s_{\theta,\hat{\lambda}}) + D(s_{\theta,\hat{\lambda}} \| s_\theta)$$

and set

$$K(r_N : s_{\theta,\hat{\lambda}}) \approx C \qquad (4.12)$$

where C is a constant independent of θ. The approximation is justified by the fact that the inaccuracy $K(r_N : s_{\theta,\hat{\lambda}})$ can be made almost independent of θ through a proper choice of $h(y, z)$, see Sect. 4.5.

By (3.85) and (3.87), we have

$$D(s_{\theta,\hat{\lambda}} \| s_\theta) = \min_{r \in \mathcal{R}_N} D(r \| s_\theta)$$

$$= \max_\lambda \left(\lambda^T \bar{h}_N - \psi(\theta, \lambda) \right).$$

Note that $D(r \| s_\theta)$ stands now for the unnormalized K-L distance of $r(y, z)$ and $s_\theta(y|z)$. The point is discussed in detail in Sect. 4.3.

Introducing the notation

$$D(\mathcal{R}_N \| s_\theta) \overset{\triangle}{=} \min_{r \in \mathcal{R}_N} D(r \| s_\theta) \qquad (4.13)$$

and substituting for $\psi(\theta, \lambda)$ from (3.76), we obtain

$$\boxed{D(\mathcal{R}_N \| s_\theta) = \max_\lambda \left(\lambda^T \bar{h}_N - \log \int s_\theta(y|z) \exp(\lambda^T h(y, z)) \right).} \qquad (4.14)$$

Hence, under (4.12) and taking into account that $\bar{K}(r_N{:}s_\theta) = K(r_N{:}s_\theta)$, we have

$$\bar{K}(r_N{:}s_\theta) \approx C + D(\mathcal{R}_N \| s_\theta). \tag{4.15}$$

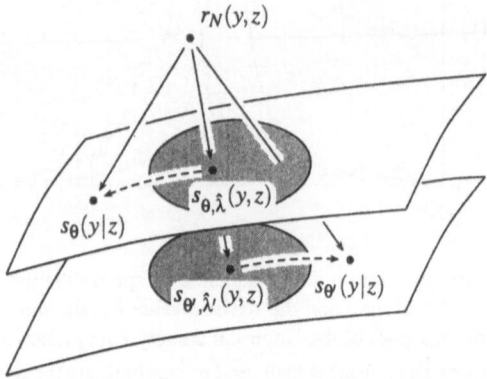

$r_N(y,z)$

$s_{\theta,\hat{\lambda}}(y,z)$

$s_\theta(y|z)$

$s_{\theta',\hat{\lambda}'}(y,z)$ $s_{\theta'}(y|z)$

Fig. 4.7. The idea of approximate estimation for dependent data. Instead of computing the unmeasurable inaccuracy $\bar{K}(r_N{:}s_\theta)$, the minimum K-L distance $D(\mathcal{R}_N \| s_\theta) = D(s_{\theta,\hat{\lambda}} \| s_\theta)$ is computed for every $\theta \in \mathcal{T}$. The projection "surfaces" correspond to the sets of all functions $C s_\theta(y|z) \exp(\lambda^T h(y,z))$ with $C > 0$. The minimum K-L distance projections $s_{\theta,\hat{\lambda}}(y,z)$ are searched for within the subsets of normalized densities of (Y, Z) (shaded area).

Approximation of Likelihood. Substituting from (4.15) for $\bar{K}(r_N{:}s_\theta)$ in (2.33), we obtain the approximate likelihood function

$$\hat{l}_N(\theta) = C \exp\big(-N D(\mathcal{R}_N \| s_\theta)\big). \tag{4.16}$$

Similarly, substituting from (4.15) for $\bar{K}(r_N{:}s_\theta)$ in (2.34), we obtain the approximate posterior density

$$\hat{p}_N(\theta) \propto p_0(\theta) \exp\big(-N D(\mathcal{R}_N \| s_\theta)\big). \tag{4.17}$$

Example 4.2 (*Linear autoregression with Cauchyian noise*)   We considered a sequence of observations $Y_1, \ldots, Y_{N+1}$ modelled through

$$Y_k = \theta Y_{k-1} + E_k$$

where $E_1, \ldots, E_{N+1}$ was a sequence of independent, Cauchy-distributed random variables with a common density

$$n(e) = \frac{1}{\pi} \frac{1}{1 + e^2}.$$

The theoretical density of Y_k given $Z_k = Y_{k-1}$ was thus

$$s_\theta(y|z) = \frac{1}{\pi} \frac{1}{1+(y-\theta z)^2}.$$

The problem was to estimate the regression coefficient θ given a particular sequence of 101 data $(y_1, \ldots, y_{101})$ shown in Fig. 4.8.

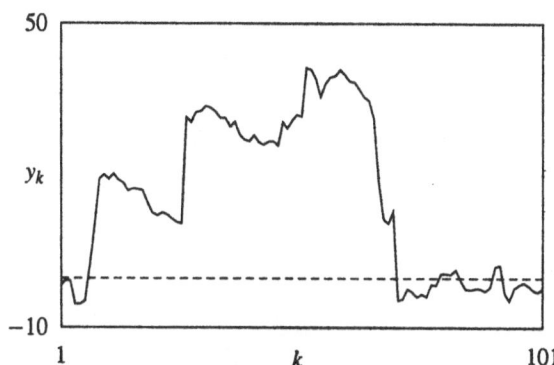

Fig. 4.8. Linear autoregression with Cauchyian noise: a sequence of 101 samples of $Y_k = 0.98\,Y_{k-1} + E_k$ with Cauchy-distributed noise $E_k \sim C(0,1)$.

We used a vector statistic $h(y)$ of dimension $n = 5$ composed of score functions again, i.e., the first-order derivatives of the log-density $\log s_\theta(y|z)$ with respect to θ

$$h_i(y,z) = \frac{2(y-\theta_i^*)}{1+(y-\theta_i^*)^2} \tag{4.18}$$

at the points $\theta_i^* = -1, -0.5, 0, +0.5, +1$. Three of these functions, h_2, h_3, h_4 are shown in Fig. 4.9.

Given the statistic $h(y,z)$ and the observed data $(y_1, \ldots, y_{101})$, we computed the sample average

$$\bar{h}_N = \frac{1}{N} \sum_{k=m+1}^{N+m} h(y_k, z_k)$$

for $N = 100$ and $m = 1$.

Given the value $\bar{h}_N$, we solved the optimization problem (4.14) for a set of different values of the unknown parameter θ, namely 121 values equidistantly located within the interval $[-2, 2]$. The resulting K-L distance $D(\mathcal{R}_N \| s_\theta)$ is shown as a solid line in the upper plot in Fig. 4.10.

To illustrate that K-L distance $D(r \| s_\theta)$ for every $r(y,z) \in \mathcal{R}_N$ is bounded from below by $D(\mathcal{R}_N \| s_\theta)$, we calculated $D(r_i \| s_\theta)$ for six such densities $r_i(y,z)$, constructed as minimum K-L distance projections of the theoretical densities $s_{\theta_i^*}(y|z)$ for $\theta_i^* = -1.5, -0.9, -0.3, +0.3, +0.9, +1.5$ onto $\mathcal{R}_N$.

The lower plot in Fig. 4.10 compares the normalized likelihoods

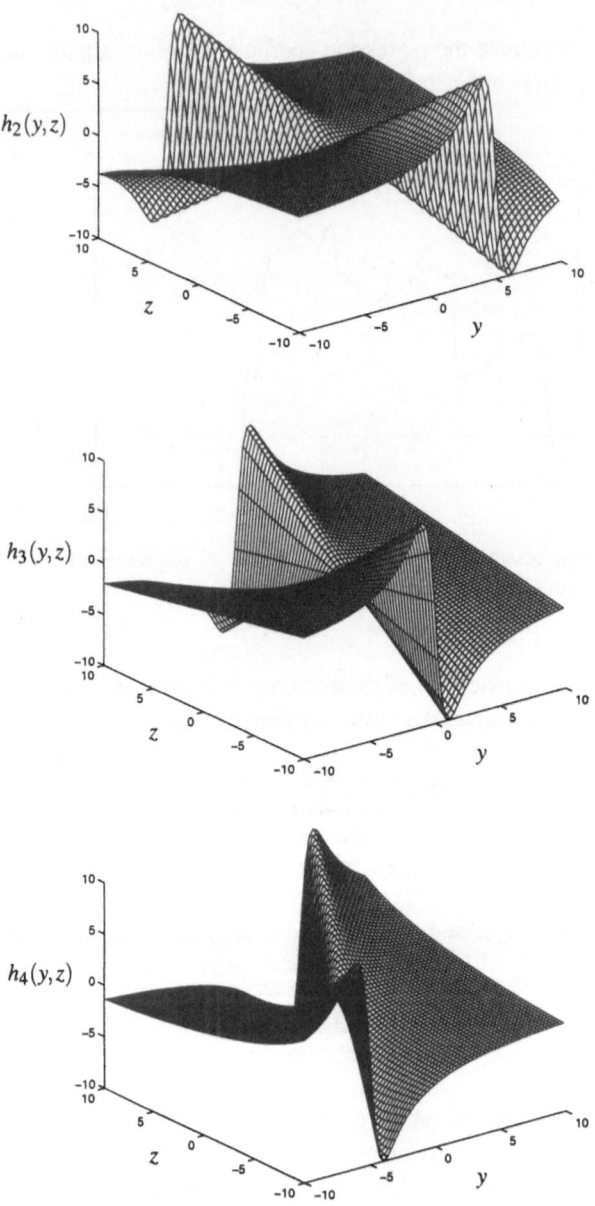

Fig. 4.9. Linear autoregression with Cauchyian noise: three of five single-data statistics $h_i(y,z)$ used for data compression.

$$\hat{l}_N(\theta) = \exp\big(-N D(\mathcal{R}_N \| s_\theta)\big),$$
$$l_N(\theta; r_i) = \exp\big(-N D(r_i \| s_\theta)\big)$$

for $N = 100$.   □

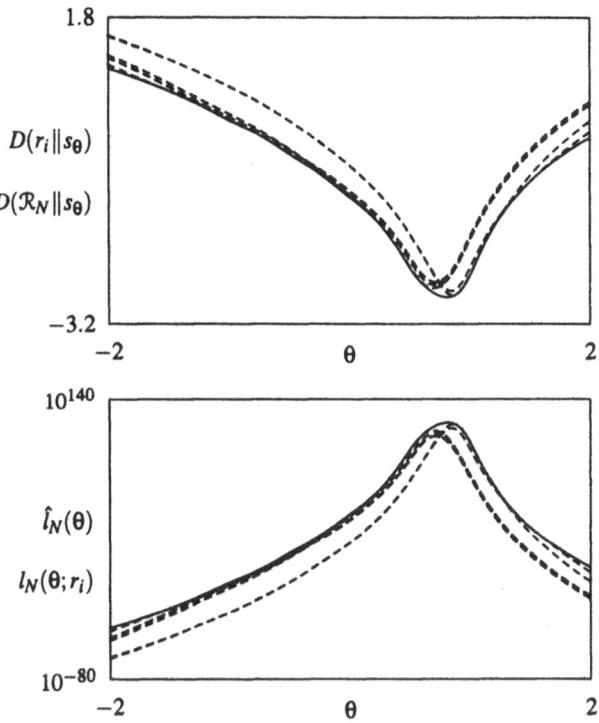

Fig. 4.10. Linear autoregression with Cauchyian noise: results of approximate estimation for $N = 100$. The upper plot compares K-L distances, the lower plot shows the normalized likelihoods.

4.3 Key Properties of Approximation

The approximations (4.8) and (4.15) are supported by a number of appealing properties.

Independent Observations

Upper and Lower Bounds. The nonnegativity (3.16) of K-L distance and the definition (4.6) of minimum K-L distance imply the following bounds

$$0 \leq D(\mathcal{R}_N \| s_\theta) \leq D(r \| s_\theta) \tag{4.19}$$

for all $r \in \mathcal{R}_N$.

ML Consistency. The minimum inaccuracy or maximum likelihood (ML) estimate

$$\min_{\theta \in \mathcal{T}} K(r_N : s_\theta)$$

naturally extends to the case of compressed data via

$$\min_{\theta \in \mathcal{T}} D(\mathcal{R}_N \| s_\theta) = \min_{\theta \in \mathcal{T}} \min_{r \in \mathcal{R}_N} D(r \| s_\theta) \tag{4.20}$$

(cf. Csiszár, 1985).

One can easily verify that if the actual distribution of data coincides with one of the theoretical distributions, the approximate estimator will always locate it. Namely, if there exists θ^* such that $\bar{h}_N = \hat{h}(\theta^*)$, then $D(\mathcal{R}_N \| s_{\theta^*}) = 0$. Hence,

$$E_{\theta^*}(h(Y)) = E_N(h(Y)) \quad \text{implies} \quad D(\mathcal{R}_N \| s_{\theta^*}) = 0 \tag{4.21}$$

and because $D(\mathcal{R}_N \| s_\theta) \geq 0$, the point θ^* is certainly among the optimum estimates.

Asymptotic ML Consistency. The above property naturally extends to the asymptotic case. If there exists θ_0 such that $\bar{h}_N \to \hat{h}(\theta_0)$ as $N \to \infty$, then $D(\mathcal{R}_N \| s_{\theta_0}) \to 0$. Hence,

$$E_N(h(Y)) \to E_{\theta_0}(h(Y)) \quad \text{implies} \quad D(\mathcal{R}_N \| s_{\theta_0}) \to 0. \tag{4.22}$$

To show it, consider a h-projection $s_{\theta, \hat{\lambda}_N}(y)$ of $r_N(y)$ onto $\mathcal{S}_{\theta_0 ; h}$. Since $\hat{h}(\theta, \hat{\lambda}_N) = \bar{h}_N$, the convergence $\bar{h}_N \to \hat{h}(\theta_0)$ implies $\hat{h}(\theta_0, \hat{\lambda}_N) \to \hat{h}(\theta_0)$. The latter implies, provided the statistics $h_0(y) \equiv 1, h_1(y), \ldots, h_n(y)$ are linearly independent, $\hat{\lambda}_N \to 0$ and $D(s_{\theta_0, \hat{\lambda}_N} \| s_{\theta_0}) \to 0$. But by (3.35), $D(s_{\theta_0, \hat{\lambda}_N} \| s_\theta) = D(\mathcal{R}_N \| s_{\theta_0})$ and thus, indeed, $D(\mathcal{R}_N \| s_{\theta_0}) \to 0$.

Taking into account that $D(\mathcal{R}_N \| s_{\theta_0}) \geq 0$, we can conclude that the approximate estimator is asymptotically consistent in the sense that the set of the optimum estimates contains the true point θ_0.

Let us emphasize that the consistency is not generally assured if the actual distribution of data does not belong to the model family considered.

Monotonicity. Owing to the convexity (3.20) of K-L distance, the minimum K-L distance $D(\mathcal{R}_N \| s_\theta)$ regarded as a function of the set argument $\mathcal{R}_N$ is (anti)monotonous in the following sense

$$\mathcal{R}_N \subseteq \mathcal{R}'_N \quad \text{implies} \quad D(\mathcal{R}_N \| s_\theta) \geq D(\mathcal{R}'_N \| s_\theta). \tag{4.23}$$

It sounds very natural: the less information we have about the data observed, the worse is our capability to discriminate between the data and model.

Relation to Likelihood. The approximation (4.8) can be given an appealing interpretation in terms of likelihoods. For this we need a slightly generalized concept of likelihood

$$l_N(s;r) = \exp(-NK(r{:}s))$$

where $r(y)$ stands for the actual density of Y while $s(y)$ is a model density of Y.

Clearly, for every $r(y) \in \mathcal{R}_N$ and every $s_\theta(y) \in \mathcal{S}$ such that $D(r\|s_\theta) < \infty$, it holds

$$K(r{:}s_\theta) = K(r{:}r) + D(r\|s_\theta)$$
$$= \min_s K(r{:}s) + D(r\|s_\theta).$$

In terms of the generalized likelihoods, the same reads

$$l_N(s_\theta;r) = \left(\max_s l_N(s;r)\right) \exp(-ND(r\|s_\theta)). \tag{4.24}$$

By the Pythagorean relationship (3.31), we have for every $r(y) \in \mathcal{R}_N$ and every $s_\theta(y) \in \mathcal{S}$

$$K(r{:}s_\theta) = K(r{:}s_{\theta,\hat{\lambda}}) + D(s_{\theta,\hat{\lambda}}\|s_\theta)$$
$$= \min_\lambda K(r{:}s_{\theta,\lambda}) + D(s_{\theta,\hat{\lambda}}\|s_\theta)$$

which implies

$$l_N(s_\theta;r) = \left(\max_\lambda l_N(s_{\theta,\lambda};r)\right) \exp(-ND(s_{\theta,\hat{\lambda}}\|s_\theta)). \tag{4.25}$$

Let us normalize the likelihood $l(s_\theta;r)$ by the maximum likelihood value that can be attained on the set of all possible models $\{s(y)\}$ and on the family $\mathcal{S}_{\theta;h}$, respectively

$$l_N(\theta;r) \triangleq \frac{l_N(s_\theta;r)}{\max_s l_N(s;r)} = \exp(-ND(r\|s_\theta)) \le 1,$$
$$\hat{l}_N(\theta) \triangleq \frac{l_N(s_\theta;r)}{\max_\lambda l_N(s_{\theta,\lambda};r)} = \exp(-ND(s_{\theta,\hat{\lambda}}\|s_\theta)) \le 1.$$

Note that we can extend the above definition of $l_N(\theta;r)$ even to $r(y) \in \mathcal{R}_N$ such that $D(r\|s_\theta) = \infty$ by setting $l_N(\theta;r) = 0$.

Since by (3.35)

$$D(s_{\theta,\hat{\lambda}}\|s_\theta) = \min_{r \in \mathcal{R}_N} D(r\|s_\theta),$$

we have in terms of the *normalized likelihoods*

$$\boxed{\hat{l}_N(\theta) = \max_{r \in \mathcal{R}_N} l_N(\theta;r).} \tag{4.26}$$

Thus, the approximate normalized likelihood $\hat{l}_N(\theta)$ can be regarded as the minimum upper bound of the normalized likelihoods $l_N(\theta;r)$ for all $r(y) \in \mathcal{R}_N$. See the lower plot in Fig. 4.5 where the property is illustrated for six $r_i(y) \in \mathcal{R}_N$.

Let us stress the importance of the normalizing divisor

$$\max_s l_N(s;r) = \exp\big(-N \min_s K(r{:}s)\big)$$
$$= \exp\big(-N K(r{:}r)\big)$$
$$= \exp\big(-N H(r)\big).$$

Without normalization, we would prefer among $r(y) \in \mathcal{R}_N$ those with *smaller* Shannon's entropy. There is no rationale for that; in fact, approximation based directly on the (unnormalized) likelihoods would make little sense.

Controlled Dynamic Systems

Unnormalized Inaccuracy. The (unnormalized) joint K-L distance $D(r\|s_\theta)$ can be decomposed as follows

$$D(r\|s_\theta) = \iint r(y,z) \log \frac{r(y,z)}{s_\theta(y|z)} \, dy \, dz$$
$$= \iint r(y,z) \log \frac{r(y,z)}{s_\theta(y|z)\,\bar{r}(z)} \, dy \, dz - \int \bar{r}(z) \log \frac{1}{\bar{r}(z)} \, dz$$

provided all the integrals exist. As a result, we have

$$\boxed{D(r\|s_\theta) = \bar{D}(r\|s_\theta) + H(\bar{r})} \tag{4.27}$$

where $\bar{r}(z) = \int r(y,z)\,dy$ stands for the marginal density of Z.

Hence, when minimizing $D(r\|s_\theta)$ over $r \in \mathcal{R}_N$, we seek a compromise between minimizing the conditional K-L distance $\bar{D}(r\|s_\theta)$ and maximizing the marginal Shannon's entropy $H(\bar{r})$. In other words, we look for a trade-off between choosing the best fit of model to data, given a particular $\bar{r}(z)$, and picking the maximum-entropy $\bar{r}(z)$ from $\mathcal{R}_N$.

Upper and Lower Bounds. Taking together the identity (4.27), the definition (4.13) of the minimum K-L distance $D(\mathcal{R}_N\|s_\theta)$ and the nonnegativity (3.18) of the conditional K-L distance $\bar{D}(r\|s_\theta)$, we get the following bounds on $D(\mathcal{R}_N\|s_\theta)$

$$\boxed{-\max_{r \in \mathcal{R}_N} H(\bar{r}) \le D(\mathcal{R}_N\|s_\theta) \le D(r\|s_\theta)} \tag{4.28}$$

for all $r(y,z) \in \mathcal{R}_N$.

Lack of ML Consistency. Let the approximate minimum inaccuracy or maximum likelihood (ML) estimate given compressed data be defined through

$$\min_{\theta \in \mathcal{T}} D(\mathcal{R}_N\|s_\theta) = \min_{\theta \in \mathcal{T}} \min_{r \in \mathcal{R}_N} D(r\|s_\theta). \tag{4.29}$$

It may happen that even if there exists θ^* such that $s_{\theta^*}(y|z)\tilde{r}(z) \in \mathcal{R}_N$, still

$$\bar{D}(s_{\theta^*} \| s_{\theta^*} | \tilde{r}) - H(\tilde{r}) > \min_{\theta \in \mathcal{T}} \min_{r \in \mathcal{R}_N} D(r \| s_\theta),$$

i.e., θ^* is not the optimum estimate. In other words, the increase of $\bar{D}(s_{\theta^*+\Delta\theta} \| s_{\theta^*} | \tilde{r}) > 0$ due to a deviation of θ from θ^* may be compensated by the increase of the marginal entropy $H(\tilde{r})$. Therefore, even if the actual conditional distribution of data coincides with a certain distribution from the model family considered, the approximate ML estimator (4.29) need not be consistent.

To ensure consistency, the statistic would have to be rich enough to satisfy

$$E_{\tilde{r}}\big(\psi(\theta,\tau;Z)\big) = E_N\big(\psi(\theta,\tau;Z)\big) \tag{4.30}$$

for all θ, τ and $r(y,z) \in \mathcal{R}_N$. By (3.47), the above implies also

$$E_{\tilde{r}}\big(\hat{f}(\theta,\tau;Z)\big) = E_N\big(\hat{f}(\theta,\tau;Z)\big)$$

for all θ, τ and $r(y,z) \in \mathcal{R}_N$. In such a case, there is an equality in (3.52) and consequently

$$\bar{D}(s_{\theta,\hat{t}} \| s_\theta | \tilde{r}) = \min_{r \in \mathcal{R}_N} \bar{D}(r \| s_\theta),$$

i.e., $s_{\theta,\hat{t}}(y|z)$ is a unique solution to the minimum conditional K-L distance problem and the minimum does *not* depend on a particular $\tilde{r}(z)$ for $r \in \mathcal{R}_N$. As a result, under the condition (4.30), we have

$$\begin{aligned}
D(\mathcal{R}_N \| s_\theta) &= \min_{r \in \mathcal{R}_N} D(r \| s_\theta) \\
&= \min_{r \in \mathcal{R}_N} \bar{D}(r \| s_\theta) - \max_{r \in \mathcal{R}_N} H(\tilde{r}).
\end{aligned}$$

Then if there exists θ^* such that $s_{\theta^*}(y|z)\tilde{r}(z) \in \mathcal{R}_N$, the minimum K-L distance (4.20) reaches indeed its absolute lower bound at $\theta = \theta^*$

$$\begin{aligned}
\min_{\theta \in \mathcal{T}} \min_{r \in \mathcal{R}_N} D(r \| s_\theta) &= \bar{D}(s_{\theta^*} \| s_{\theta^*} | \tilde{r}) - \max_{r \in \mathcal{R}_N} H(\tilde{r}) \\
&= -\max_{r \in \mathcal{R}_N} H(\tilde{r}),
\end{aligned}$$

i.e., the approximate ML estimator is consistent in this case.

We stress again that the consistency is not generally assured if the actual distribution of data does not belong to the model family considered.

The condition (4.30) is very strict and rarely met with practically interesting models except the linear normal ones. As a rule, it can be satisfied only approximately for a certain region $A \subset \mathcal{T}$ in the parameter space. However disappointing it looks, it is quite a natural result; without being able to compute the empirical expectations

$$E_N\big(\hat{f}(\theta,\tau;Z)\big)$$

for all θ and τ, we cannot hope for perfect fit between data and model in general.

Monotonicity. It is easy to check that the convexity (3.20) of K-L distance is preserved even when the second argument $s_\theta(y|z)$ is not normalized with respect to z. The minimum K-L distance $D(\mathcal{R}_N \| s_\theta)$ regarded as a function of the set argument $\mathcal{R}_N$ is thus again (anti)monotonous in the sense that

$$\boxed{\mathcal{R}_N \subseteq \mathcal{R}'_N \quad \text{implies} \quad D(\mathcal{R}_N \| s_\theta) \geq D(\mathcal{R}'_N \| s_\theta).} \tag{4.31}$$

Relation to Likelihood. Similarly as in the case of independent observations, the approximation (4.15) can be given an interesting interpretation in terms of likelihoods.

We introduce a slightly generalized concept of likelihood

$$l_N(s;r) = \Gamma_N \exp(-NK(r{:}s))$$

where $r(y,z)$ stands for the actual joint density of (Y,Z) while $s(y|z)$ is a model conditional density of Y given Z. The factor Γ_N is independent of s. Remember that $\Gamma_N = 1$ for uncontrolled dynamic systems and

$$\Gamma_N = \prod_{k=m+1}^{N+m} \gamma_k(u_k|y^{k-1}, u^{k-1})$$

for controlled dynamic systems.

For every $r(y,z) \in \mathcal{R}_N$ and every $s_\theta(y|z) \in \mathcal{S}$ such that $D(r \| s_\theta) < \infty$, it holds

$$\begin{aligned} K(r{:}s_\theta) &= K(r{:}r) + D(r \| s_\theta) \\ &= \min_s K(r{:}s) + D(r \| s_\theta), \end{aligned}$$

i.e., in terms of the generalized likelihoods

$$l_N(s_\theta;r) = \left(\max_s l_N(s;r) \right) \exp(-ND(r \| s_\theta)). \tag{4.32}$$

By the Pythagorean relationship (3.81), we have for every $r(y,z) \in \mathcal{R}_N$ and every $s_\theta(y|z) \in \mathcal{S}$

$$\begin{aligned} K(r{:}s_\theta) &= K(r{:}s_{\theta,\hat{\lambda}}) + D(s_{\theta,\hat{\lambda}} \| s_\theta) \\ &= \min_\lambda K(r{:}s_{\theta,\lambda}) + D(s_{\theta,\hat{\lambda}} \| s_\theta) \end{aligned}$$

which implies

$$l_N(s_\theta;r) = \left(\max_\lambda l_N(s_{\theta,\lambda};r) \right) \exp(-ND(s_{\theta,\hat{\lambda}} \| s_\theta)). \tag{4.33}$$

We normalize the likelihood $l(s_\theta;r)$ by the maximum likelihood value on the set of all possible models $\{s(y|z)\}$ and on the family $\mathcal{S}_{\theta;h}$, respectively

$$l_N(\theta;r) \triangleq \frac{l_N(s_\theta;r)}{\max_s l_N(s;r)} = \exp(-ND(r \| s_\theta)),$$

$$\hat{l}_N(\theta) \triangleq \frac{l_N(s_\theta;r)}{\max_\lambda l_N(s_{\theta,\lambda};r)} = \exp(-ND(s_{\theta,\hat{\lambda}} \| s_\theta)).$$

Again, we can extend the definition of $l_N(\theta;r)$ even to $r(y,z) \in \mathcal{R}_N$ such that $D(r\|s_\theta) = \infty$ by setting $l_N(\theta;r) = 0$. Note that the normalized likelihoods are not bounded from above by 1 as they are in the case of independent observations.

Since by (3.85)

$$D(s_{\theta,\hat{\lambda}}\|s_\theta) = \min_{r \in \mathcal{R}_N} D(r\|s_\theta),$$

we have in terms of the *normalized likelihoods*

$$\boxed{\hat{l}_N(\theta) = \max_{r \in \mathcal{R}_N} l_N(\theta;r).} \tag{4.34}$$

The approximate normalized likelihood $\hat{l}_N(\theta)$ can thus be constructed as the minimum upper bound of the normalized likelihoods $l_N(\theta;r)$ for all $r(y,z) \in \mathcal{R}_N$. See the lower plot in Fig. 4.10 where the property is illustrated for six $r_i(y,z) \in \mathcal{R}_N$.

Normalization by the factor

$$\begin{aligned}
\max_s l_N(s;r) &= \Gamma_N \exp\big(-N \min_s K(r{:}s)\big) \\
&= \Gamma_N \exp\big(-N K(r{:}r)\big) \\
&= \Gamma_N \exp\big(-N H(r)\big)
\end{aligned}$$

means that we get rid of all factors that cannot be affected by the choice of the model density $s(y|z)$.

4.4 Relation to Large Deviation Theory

This section indicates existing connections between the approximation proposed in Sect. 4.2 and an asymptotic approximation of the joint probability of sample. The results are stated here without proofs but with references to the original sources. For simplicity, the random variables $Y_1, Y_2, \ldots$ are considered discrete only, taking values in a finite set $\mathcal{Y}$.

Independent Observations

Large Deviation Asymptotics. Consider a sequence $(Y_1, \ldots, Y_N)$ of independent and identically distributed random variables with the empirical mass function $r_N(y)$. The random variables Y are distributed according to a common probability mass function $s_\theta(y)$ possibly parametrized by an unknown parameter $\theta \in \mathcal{T}$. We show that the probability that the empirical mass function $r_N(y)$ belongs to a certain set $\mathcal{R}$

$$q_\theta^N(\{y^N : r_N \in \mathcal{R}\})$$

can be approximated in the following asymptotic sense.

We say that two sequences f_N, g_N are *equal to the first order in the exponent*, $f_N \doteq g_N$, if

$$\lim_{N \to \infty} \frac{1}{N} \log \frac{f_N}{g_N} = 0.$$

By combinatorial arguments, the number of sequences y^N having the same empirical mass function $r_N(y)$ is

$$\left| \{ y^N : r_N = r \} \right| \doteq \exp(NH(r))$$

(Csiszár and Körner, 1981, Lemma 2.3) where $|A|$ denotes the number of elements of a finite set A. Since by (2.22) and (2.45)

$$
\begin{aligned}
q_\theta^N(y^N) &= \exp(-NK(r_N;s_\theta)) \\
&= \exp(-NH(r_N)) \exp(-ND(r_N \| s_\theta)),
\end{aligned}
$$

the probability of $\{ y^N : r_N = r \}$ is

$$q_\theta^N(\{ y^N : r_N = r \}) \doteq \exp(-ND(r \| s_\theta))$$

(Csiszár and Körner, 1981, Lemma 2.6). The probability $q_\theta^N(\{ y^N : r_N \in \mathcal{R} \})$ is thus

$$q_\theta^N(\{ y^N : r_N \in \mathcal{R} \}) \doteq \sum_{r \in \mathcal{R}} \exp(-ND(r \| s_\theta)).$$

The number of different empirical mass functions of all possible sequences $y^N \in \mathcal{Y}^N$ is less than $(N+1)^{|\mathcal{Y}|}$ since for every $y \in \mathcal{Y}$, $N r_N(y)$ can take on values $0, 1, \ldots, N$ only (Csiszár and Körner, 1981, Lemma 2.2). The number of terms in the above sum is thus "only" polynomial in the length N of the sequences y^N. Therefore, the sum is (under additional topological assumptions on the set $\mathcal{R}$) equal to the first order in the exponent to its largest term

$$
\begin{aligned}
q_\theta^N(\{ y^N : r_N \in \mathcal{R} \}) &\doteq \max_{r \in \mathcal{R}} \exp(-ND(r \| s_\theta)) \\
&\doteq \exp(-N \min_{r \in \mathcal{R}} D(r \| s_\theta)).
\end{aligned}
$$

More precisely, if $\mathcal{R}$ is the closure of its interior, then

$$\boxed{\lim_{N \to \infty} \frac{1}{N} \log q_\theta^N(\{ y^N : r_N \in \mathcal{R} \}) = -D(\mathcal{R} \| s_\theta)} \qquad (4.35)$$

with $D(\mathcal{R} \| s_\theta)$ defined as

$$D(\mathcal{R} \| s_\theta) \overset{\triangle}{=} \min_{r \in \mathcal{R}} D(r \| s_\theta). \qquad (4.36)$$

The result is known as Sanov's *large deviation theorem* (Sanov, 1957). The derivation outlined informally above is due to Csiszár and Körner (1981), cf. Csiszár et al. (1987) and Cover and Thomas (1991).

The probability that the empirical mass function belongs to a set that does not contain the true mass function is known (by the law of large numbers) to converge to zero. The large deviation theorem refines this statement, showing that the probability converges to zero *exponentially fast*, with the *rate* given by the Kullback-Leibler distance $D(\mathcal{R} \| s_\theta)$ between a given set $\mathcal{R}$ and the sampling mass function s_θ.

The identical structure of the large-deviation approximation (4.35)

$$q_\theta^N\left(\{y^N : r_N \in \mathcal{R}\}\right) \doteq \exp\left(-ND(\mathcal{R}\|s_\theta)\right)$$

and the Pythagorean approximation (4.9)

$$\hat{l}_N(\theta) = C \exp\left(-ND(\mathcal{R}_N\|s_\theta)\right)$$

suggests that the approximate posterior density (4.10)

$$\hat{p}_N(\theta) \propto p_0(\theta) \exp\left(-ND(\mathcal{R}_N\|s_\theta)\right)$$

is equal to the first order in the exponent to the density of Θ conditional on the statistic value

$$p(\theta|T_N(y^N) = \bar{h}_N).$$

Minimum K-L Distance. When the set $\mathcal{R}$ is more specific, more can be said about the minimum Kullback-Leibler distance in (4.36). Often $\mathcal{R}$ is supposed to be bounded by hyperplanes, i.e.,

$$\mathcal{R} = \left\{r(y) : \sum_{y \in \mathcal{Y}} r(y)\,h(y) \geq \bar{h}\right\} \tag{4.37}$$

where $h: \mathcal{Y} \to \mathbb{R}^n$ is a given vector statistic.

The minimum Kullback-Leibler distance can be found then by solving

$$D(\mathcal{R}\|s_\theta) = \max_{\lambda \geq 0}\left(\lambda^T\bar{h} - \log\sum_{y \in \mathcal{Y}} s_\theta(y)\exp(\lambda^T h(y))\right) \tag{4.38}$$

where λ is a column vector of dimension n.

The maximization (4.38) is basically a dual problem to the convex programming problem (4.36). For proof see Kullback (1959) and Csiszár (1984).

Minimum K-L Distance Projection. Given (4.37), the minimum K-L distance problem (4.36) is solved by

$$s_{\theta,\hat{\lambda}}(y) = s_\theta(y)\exp\left(\hat{\lambda}^T h(y) - \psi(\theta,\hat{\lambda})\right) \tag{4.39}$$

with $\hat{\lambda}$ satisfying

$$\sum_{y \in \mathcal{Y}} s_{\theta,\hat{\lambda}}(y)\,h(y) \geq \bar{h}. \tag{4.40}$$

The form (4.39) can be found using the Lagrange multiplier method, see Kullback (1959) and Cover and Thomas (1991).

Markov Chains

Large Deviation Asymptotics. Consider a sequence $(Y_1, \ldots, Y_{N+1})$ that forms a Markov chain with the conditional mass function $s(y|z)$, $z_k = y_{k-1}$. Let $s_\theta(y|z) > 0$ for all $(y, z) \in \mathcal{Y}^2$. Suppose that the empirical mass function $r_N(y, z)$ belongs to a certain set $\mathcal{R}$. Then the probability

$$q_\theta^N(\{y_2^{N+1} : r_N \in \mathcal{R}\}|y_1)$$

can be approximated in the following asymptotic sense.

Let $\mathcal{C}$ be the set of all mass functions $r(y, z)$ on $\mathcal{Y}^2$ such that their both marginals coincide, i.e.,

$$\sum_{z \in \mathcal{Y}} r(y, z) = \sum_{z \in \mathcal{Y}} r(z, y)$$

for all $y \in \mathcal{Y}$. Hence, any distribution from $\mathcal{C}$ determines a stationary Markov chain. If $\mathcal{R}$ is the closure of its interior, then

$$\boxed{\lim_{N \to \infty} \frac{1}{N} \log q_\theta^N(\{y_2^{N+1} : r_N \in \mathcal{R}\}|y_1) = -\bar{D}(\mathcal{R} \cap \mathcal{C} \| s_\theta)} \qquad (4.41)$$

where

$$\bar{D}(\mathcal{R} \cap \mathcal{C} \| s_\theta) = \min_{r \in \mathcal{R} \cap \mathcal{C}} \bar{D}(r \| s_\theta). \qquad (4.42)$$

For proof see Boza (1971) and Natarajan (1985). Csiszár et al. (1987) refrained (under additional regularity assumptions) from the strict positivity of s_θ; this is essential for generalization to higher-order Markov chains.

Minimum K-L Distance. Suppose that the set $\mathcal{R}$ is bounded by hyperplanes, i.e.,

$$\mathcal{R} = \left\{ r(y, z) : \sum_{(y,z) \in \mathcal{Y}^2} r(y, z) h(y, z) \geq \bar{h} \right\} \qquad (4.43)$$

where $h : \mathcal{Y}^2 \to \mathbb{R}^n$ is a given vector statistic.

The minimum conditional K-L distance is then given by

$$\boxed{\bar{D}(\mathcal{R} \cap \mathcal{C} \| s_\theta) = \max_{\lambda \geq 0} \left(\lambda^T \bar{h} - \log \omega(\theta, \lambda) \right)} \qquad (4.44)$$

where λ is a column vector of dimension n and $\omega(\theta, \lambda)$ stands for the largest eigenvalue of the square matrix of order $|\mathcal{Y}|$

$$M_{\theta, \lambda}(y, z) = s_\theta(y|z) \exp\left(\lambda^T h(y, z) \right). \qquad (4.45)$$

A simpler result for $s(y|z) = \text{const.}$ was proved in Spitzer (1972) and Justesen and Høholdt (1984). For the above expression see Csiszár et al. (1987).

Minimum K-L Distance Mass Function. Given (4.43), the minimum conditional K-L distance problem (4.42) is solved by the probability mass function

$$s_{\theta,\hat{\lambda}}(y,z) = \frac{v_{\theta,\hat{\lambda}}(y)\, M_{\theta,\hat{\lambda}}(y,z)\, w_{\theta,\hat{\lambda}}(z)}{\omega(\theta,\hat{\lambda})} \tag{4.46}$$

where $\omega(\theta,\hat{\lambda})$ is the largest eigenvalue and $v_{\theta,\hat{\lambda}}$ and $w_{\theta,\hat{\lambda}}$ are the corresponding left and right eigenvectors, normalized to have inner product 1, of the matrix $M_{\theta,\hat{\lambda}}$ (4.45) for $\hat{\lambda}$ attaining the maximum in (4.44).

An analogous result for $s(y|z) = $ const. was proved again in Spitzer (1972) and Justesen and Høholdt (1984). The above result is due to Csiszár et al. (1987).

Controlled Dynamic Systems

The large deviation approximation in the case of Markov chains copes with the uncertainty about $\tilde{r}_N(z)$ by constructing a joint probability mass function $s_{\theta,\hat{\lambda}}(y,z)$ that belongs to $\mathcal{R}$ and corresponds to a certain *stationary* Markov chain. Loosely speaking, the missing information about the distribution of Z is replaced by information that is valid asymptotically given a particular model.

There is no similar result for the case of controlled systems. The reason is obvious; as we say nothing about the probability distribution of external inputs, we do not have enough information to reconstruct the asymptotic distribution of both Y and Z given a particular model. Note that the approach we used in Chap. 3 does not expect such information, rather the distribution of Z is approximated explicitly using available information.

4.5 Choice of Statistic

The sufficient statistics for practically interesting models have often very large or even infinite dimension. To make estimation for such models feasible, we have to use a statistic of limited dimension—not sufficient for restoration of the true likelihood. The choice of the statistic seriously affects the resulting discrepancy between the true and approximate likelihoods. In the following we present a class of statistics which are the next to try if the sufficient statistics cannot be used.

Independent Observations

Necessary Statistic. Consider a family of distributions $\mathcal{S} = \{s_\theta(y) : \theta \in \mathcal{T}\}$. A statistic

$$T_N \colon \mathcal{Y}^N \to \mathbb{R}^n$$

is called *necessary* for $\mathcal{S}$ if $T_N(Y^N)$ is a function of any sufficient statistic $T_N^*(Y^N)$. Thus, the necessary statistic is a function of a minimal sufficient statistic.

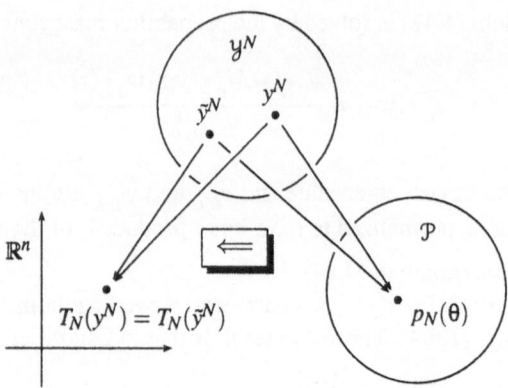

Fig. 4.11. The Bayesian view of necessary statistic for independent observations. The statistic T_N is necessary for the parametric family S if for every y^N and $\tilde{y}^N$ resulting in the same posteriors $p(\theta|y^N) = p(\theta|\tilde{y}^N)$, the corresponding values of the statistic coincide, $T_N(y^N) = T_N(\tilde{y}^N)$.

Under our regularity assumptions, the necessary statistic can be constructed as follows. Consider the linear space $\mathcal{H}$ spanned by constants and the functions

$$\log s_\theta(y) - \log s_{\theta_0}(y)$$

for all $\theta \in \mathcal{T}$ where θ_0 is an arbitrary fixed point in $\mathcal{T}$. Now choose n linearly independent, non-constant functions $h_1(y), \ldots, h_n(y)$ from $\mathcal{H}$. The functions

$$h_0(y) \equiv 1, h_1(y), \ldots, h_n(y)$$

span an $(n+1)$-dimensional linear subspace $\mathcal{H}_0$ of $\mathcal{H}$. Given the vector statistic of a single observation y

$$h(y) = [h_1(y), \ldots, h_n(y)]^T,$$

we define a vector statistic of the whole sample y^N through the empirical expectation

$$T_N(y^N) \stackrel{\triangle}{=} E_N(h(Y)) = \frac{1}{N} \sum_{k=1}^{N} h(y_k). \tag{4.47}$$

Since the empirical expectation of any set of basis vectors of the linear space $\mathcal{H}$ forms a minimal sufficient statistic, a statistic defined as the empirical expectation of basis vectors of a linear subspace $\mathcal{H}_0$ of $\mathcal{H}$ is clearly necessary.

Note that the single-data statistics $h_i(y)$, $i = 1, \ldots, n$ can be regarded as specific, model-based transformations of the observed data (see Fig. 4.12).

Construction of Single-Data Statistic. The conceptual construction of $h(y)$ shown above starts from the linear space $\mathcal{H}$ which is typically of high or even infinite dimension. For practical purposes, it is more convenient to have a more direct way of constructing directly a set of linearly independent functions from $\mathcal{H}$.

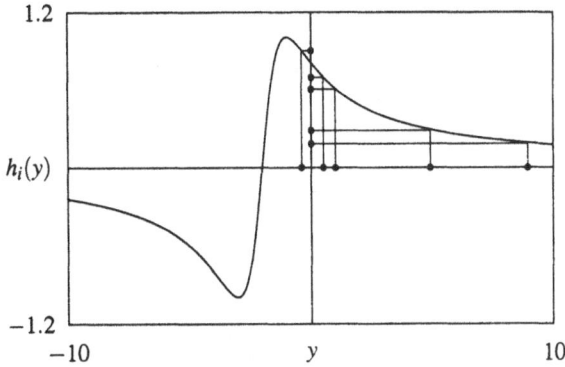

Fig. 4.12. A single-data statistic $h_i(y)$ defines a model-specific transformation of the observed data $y_1, \ldots, y_N$.

What was said in Sect. 3.5 about construction of a canonical statistic of the enveloping exponential family applies directly to our case with the only distinction—the functions $h_i(y)$, $i = 1, \ldots, n$ are not required now to form a basis of the whole linear space $\mathcal{H}$. Some typical constructions of single-data statistics follow.

Differencing. Pick up $n + 1$ points $\theta_1^*, \ldots, \theta_{n+1}^*$ in the parameter space $\mathcal{T}$ and set

$$h_i(y) = \log s_{\theta_{i+1}^*}(y) - \log s_{\theta_i^*}(y). \tag{4.48}$$

Differentiation. Suppose that $\log s_\theta(y)$ is differentiable at every $\theta \in \mathcal{T}$ and for all $y \in \mathcal{Y}$. Pick up n points $\theta_1^*, \ldots, \theta_n^*$ in the parameter space $\mathcal{T}$ and n column vectors ω_1^*, $\ldots, \omega_n^*$ from $\mathbb{R}^{\dim \theta}$. Set

$$h_i(y) = {\omega_i^*}^T \nabla_\theta \log s_{\theta_i^*}(y). \tag{4.49}$$

Weighted Integration. Pick up n weighting functions $w_1^*(\theta), \ldots, w_n^*(\theta)$ such that

$$\int w_i^*(\theta) \, d\theta = 0, \quad i = 1, \ldots, n.$$

Then set

$$h_i(y) = \int w_i^*(\theta) \log s_\theta(y) \, d\theta. \tag{4.50}$$

General Construction of Single-Data Statistic. Again, other constructions might be used like higher-order differencing, higher-order differentiation, various combinations of differentiation and weighted integration, etc. All these constructions are just special instances of the following general construction.

Consider a vector space $\tilde{\mathcal{H}}$ that contains functions

$$\tilde{h}(\theta) = \log s_\theta(y)$$

for all $\theta \in \mathcal{T}$. Let L_i, $i = 1,\ldots,n$ be a set of linear functionals defined on the vector space $\tilde{\mathcal{H}}$. Suppose in addition that the linear functionals are normalized so that

$$L_i(1) = 0$$

for $i = 1,\ldots,n$. Then define

$$\boxed{h_i(y) = L_i\big(\log s_\theta(y)\big)} \qquad (4.51)$$

for $i = 1,\ldots,n$.

When the functions $h_0(y) \equiv 1$, $h_1(y)$, $\ldots$, $h_n(y)$ are linearly independent, the vector function $h(y) = [h_1(y),\ldots,h_n(y)]^T$ forms a single-data vector statistic that defines through (4.47) an n-dimensional statistic necessary for the family $\mathcal{S}$.

Interpretation of Necessary Statistic. With respect to the connection between the empirical expectation of the log-density $\log s_\theta(Y)$ and the log-likelihood $\log l_N(\theta)$

$$E_N\big(\log s_\theta(Y)\big) = \frac{1}{N}\log l_N(\theta),$$

the empirical expectation of the single-data statistics (4.48), (4.49), (4.50) gives

$$E_N\left(\log \frac{s_{\theta_{i+1}^*}(Y)}{s_{\theta_i^*}(Y)}\right) = \frac{1}{N}\log \frac{l_N(\theta_{i+1}^*)}{l_N(\theta_i^*)},$$

$$E_N\left(\omega_i^{*T}\nabla_\theta \log s_{\theta_i^*}(Y)\right) = \frac{1}{N}\omega_i^{*T}\nabla_\theta \log l_N(\theta_i^*),$$

$$E_N\left(\int w_i^*(\theta)\log s_\theta(Y)\,d\theta\right) = \frac{1}{N}\int w_i^*(\theta)\log l_N(\theta)\,d\theta,$$

respectively.

In general, we have by (4.51)

$$E_N\big(L_i(\log s_\theta(Y))\big) = \frac{1}{N}L_i\big(\log l_N(\theta)\big).$$

Similarly as a minimal sufficient statistic, the necessary statistic provides condensed information about the "shape" of likelihood $\log l_N(\theta)$, in general not sufficient, however, for its complete restoration.

Likelihood Matching. Applying the linear functionals $L_i(\cdot)$, $i = 1,\ldots,n$ to both sides of the Pythagorean relationship (3.31)

$$K(r_N{:}s_\theta) = K(r_N{:}s_{\theta,\hat{\lambda}}) + D(s_{\theta,\hat{\lambda}}\|s_\theta)$$

and taking into account (2.23) and (4.9)

$$L_i\big(K(r_N{:}s_\theta)\big) = L_i\Big(-\frac{1}{N}\log l_N(\theta)\Big),$$

$$L_i\big(D(s_{\theta,\hat\lambda}\|s_\theta)\big) = L_i\Big(-\frac{1}{N}\log \hat l_N(\theta)\Big),$$

we have

$$L_i\Big(-\frac{1}{N}\log l_N(\theta)\Big) = L_i\big(K(r_N{:}s_{\theta,\hat\lambda})\big) + L_i\Big(-\frac{1}{N}\log \hat l_N(\theta)\Big). \tag{4.52}$$

Note that $L_i\big(K(r_N{:}s_\theta)\big)$ essentially measures sensitivity of $K(r_N{:}s_\theta)$ to a specific change of the model density $s_\theta(y)$, namely

$$
\begin{aligned}
L_i\big(K(r_N{:}s_\theta)\big) &= L_i\Big(\int r_N(y)\log\frac{1}{s_\theta(y)}\,dy\Big)\\
&= \int r_N(y)\log\frac{1}{\exp\big(L_i(\log s_\theta(y))\big)}\,dy\\
&= \int r_N(y)\log\frac{1}{\exp\big(h_i(y)\big)}\,dy\\
&= \frac{d}{d\mu}\int r_N(y)\log\frac{1}{s_\theta(y)\exp(\mu h_i(y))}\,dy\\
&= \frac{d}{d\mu}K\big(r_N{:}s_\theta\exp(\mu h_i)\big)
\end{aligned}
$$

where we used $h_i(y) = L_i\big(\log s_\theta(y)\big)$. The statistic defined through $h(y)$ is certainly sufficient for this kind of change so that

$$L_i\big(K(r_N{:}s_{\theta,\hat\lambda})\big) = 0. \tag{4.53}$$

To put it a different way, the projection $s_{\theta,\hat\lambda}(y)$ of $s_\theta(y)$ onto $\mathcal{R}_N$ does not change under the above change.

Taking (4.52) and (4.53) together, we obtain after straightforward manipulations

$$\boxed{L_i\big(\log l_N(\theta)\big) = L_i\big(\log \hat l_N(\theta)\big)} \tag{4.54}$$

for $i = 1,\dots,n$. Thus, even if the necessary statistic does not bring enough information for restoration of a complete log-likelihood $\log l_N(\theta)$, a perfect matching between the true and approximate log-likelihoods is guaranteed at least locally through (4.54).

For instance, when the single-data statistic is constructed through (4.48), the true and approximate log-likelihoods coincide, up to an additive constant, on a fixed grid of points $\theta_1^*, \dots, \theta_{n+1}^*$ in the parameter space $\mathcal{T}$. This is, by the way, the essence of the well-known point-mass approximation introduced by Bucy (1969) and Bucy and Senne (1971). The approximation proposed in Sect. 4.2 can be regarded in this respect as an advanced, model-based interpolation between the grid points.

Similarly, when the single-data statistic is constructed through (4.49), the true and approximate likelihoods have the same (directional) derivatives on a fixed grid of points $\theta_1^*, \dots, \theta_n^*$ in the parameter space $\mathcal{T}$ (cf. Fig. 4.13).

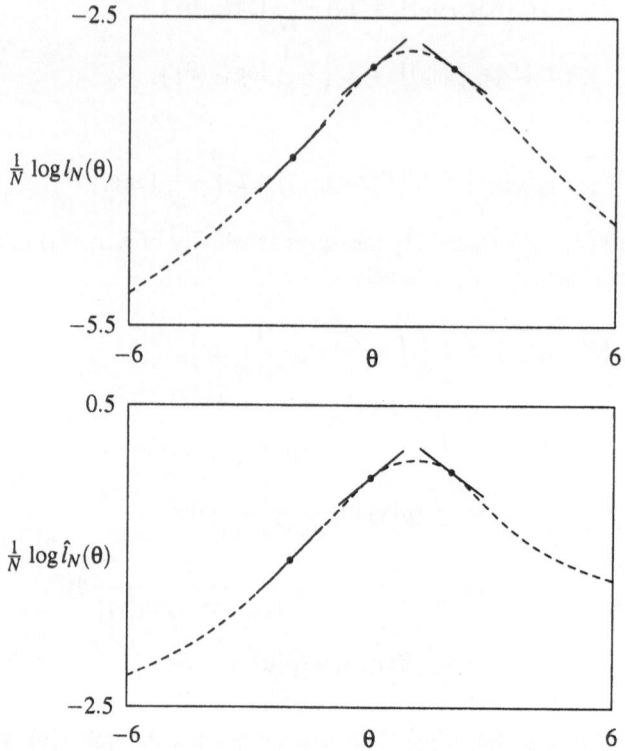

Fig. 4.13. Cauchyian noise: owing to the choice of the single-data statistics $h_i(y)$, $i = 1,2,3$ as score functions at $\theta_i^* = -2,0,+2$, the derivatives of the true (upper) and approximate (lower) normalized log-likelihoods agree at those points.

Controlled Dynamic Systems

Necessary Statistic. Consider a family of distributions $\mathcal{S} = \{s_\theta(y|z) : \theta \in \mathcal{T}\}$. A statistic

$$T_N : \mathcal{Y}^{N+m} \times \mathcal{U}^{N+m} \to \mathbb{R}^n$$

is said to be *necessary* for $\mathcal{S}$ if $T_N(Y^{N+m}, U^{N+m})$ is a function of any sufficient statistic $T_N^*(Y^{N+m}, U^{N+m})$. Thus, the necessary statistic is a function of a minimal sufficient statistic.

Assuming the regularity conditions satisfied, the necessary statistic can be constructed as follows. Consider the linear space $\mathcal{H}$ spanned by constants and the functions

$$\log s_\theta(y|z) - \log s_{\theta_0}(y|z)$$

for all $\theta \in \mathcal{T}$ where θ_0 is an arbitrary fixed point in $\mathcal{T}$. Pick up n linearly independent, non-constant functions $h_1(y,z), \dots, h_n(y,z)$ from $\mathcal{H}$. The functions

$$h_0(y,z) \equiv 1, h_1(y,z), \dots, h_n(y,z)$$

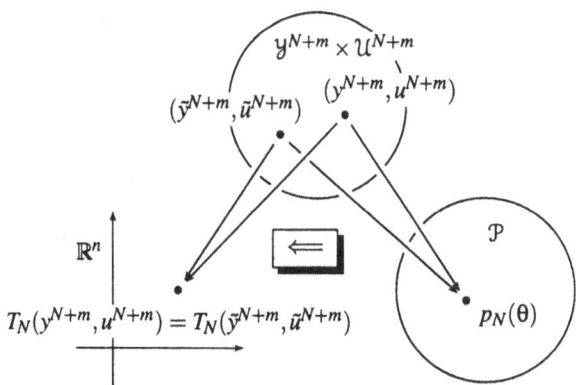

Fig. 4.14. The Bayesian view of necessary statistic for controlled dynamic systems. The statistic T_N is necessary for the parametric family S if for every (y^{N+m}, u^{N+m}) and $(\bar{y}^{N+m}, \bar{u}^{N+m})$ resulting in the same posterior densities $p(\theta|\bar{y}^{N+m}, \bar{u}^{N+m}) = p(\theta|y^{N+m}, u^{N+m})$ the corresponding values of the statistic coincide, $T_N(\bar{y}^{N+m}, \bar{u}^{N+m}) = T_N(y^{N+m}, u^{N+m})$.

span an $(n+1)$-dimensional linear subspace $\mathcal{H}_0$ of $\mathcal{H}$. Given the vector statistic of single observation (y, z)

$$h(y,z) = [h_1(y,z), \dots, h_n(y,z)]^T,$$

we define a vector statistic of the whole sample (y^{N+m}, u^{N+m}) as the empirical expectation or sample average

$$T_N(y^{N+m}, u^{N+m}) \triangleq E_N\big(h(Y,Z)\big) = \frac{1}{N} \sum_{k=m+1}^{N+m} h(y_k, z_k). \qquad (4.55)$$

As the empirical expectation of any set of basis vectors of the linear space $\mathcal{H}$ is a minimal sufficient statistic, a statistic defined through the empirical expectation of basis vectors of a linear subspace $\mathcal{H}_0$ of $\mathcal{H}$ is necessary.

Construction of Single-Data Statistic. Similarly as for independent observations, there are many possible constructions of the vector statistic $h(y, z)$. The following are perhaps most typical.

Differencing. Pick up $n+1$ points $\theta_1^*, \dots, \theta_{n+1}^*$ in the parameter space $\mathcal{T}$ and set

$$h_i(y,z) = \log s_{\theta_{i+1}^*}(y|z) - \log s_{\theta_i^*}(y|z). \qquad (4.56)$$

Differentiation. Suppose that $\log s_\theta(y|z)$ is differentiable at every $\theta \in \mathcal{T}$ and for all $(y, z) \in \mathcal{Y} \times \mathcal{Z}$. Pick up n points $\theta_1^*, \dots, \theta_n^*$ in the parameter space $\mathcal{T}$ and n vectors $\omega_1^*, \dots, \omega_n^*$ from $\mathbb{R}^{\dim \theta}$. Set

$$h_i(y,z) = \omega_i^{*T} \nabla_\theta \log s_{\theta_i^*}(y|z). \tag{4.57}$$

Weighted Integration. Pick up n weighting functions $w_1^*(\theta), \ldots, w_n^*(\theta)$ such that

$$\int w_i^*(\theta)\, d\theta = 0, \quad i = 1, \ldots, n$$

and set

$$h_i(y,z) = \int w_i^*(\theta) \log s_\theta(y|z)\, d\theta. \tag{4.58}$$

General Construction of Single-Data Statistic. A general construction looks as follows. Consider a vector space $\mathcal{H}$ that contains functions

$$\tilde{h}(\theta) = \log s_\theta(y|z)$$

for all $\theta \in \mathcal{T}$. Let L_i, $i = 1, \ldots, n$ be a set of linear functionals defined on the vector space $\mathcal{H}$. Suppose in addition that the linear functionals are normalized so that

$$L_i(1) = 0$$

for $i = 1, \ldots, n$. Then define

$$h_i(y,z) = L_i\big(\log s_\theta(y|z)\big) \tag{4.59}$$

for $i = 1, \ldots, n$.

When the functions $h_0(y,z) \equiv 1, h_1(y,z), \ldots, h_n(y,z)$ are linearly independent, the vector function $h(y,z) = [h_1(y,z), \ldots, h_n(y,z)]^T$ forms a single-data vector statistic that defines through (4.55) an n-dimensional statistic necessary for the family $\mathcal{S}$.

Interpretation of Necessary Statistic. The picture is perfectly analogous to the case of independent observation.

Taking into account the connection between the empirical expectation of the log-density $\log s_\theta(Y|Z)$ and the log-likelihood $\log l_N(\theta)$

$$E_N\big(\log s_\theta(Y|Z)\big) = C + \frac{1}{N} \log l_N(\theta),$$

the empirical expectation of the single-data statistics (4.56), (4.57), (4.58) yields

$$E_N\left(\log \frac{s_{\theta_{i+1}^*}(Y|Z)}{s_{\theta_i^*}(Y|Z)}\right) = \frac{1}{N} \log \frac{l_N(\theta_{i+1}^*)}{l_N(\theta_i^*)},$$

$$E_N\left(\omega_i^{*T} \nabla_\theta \log s_{\theta_i^*}(Y|Z)\right) = \frac{1}{N} \omega_i^{*T} \nabla_\theta \log l_N(\theta_i^*),$$

$$E_N\left(\int w_i^*(\theta) \log s_\theta(Y|Z)\, d\theta\right) = \frac{1}{N} \int w_i^*(\theta) \log l_N(\theta)\, d\theta,$$

respectively.

In general, we have by (4.59)

$$E_N\left(L_i\left(\log s_\theta(Y|Z)\right)\right) = \frac{1}{N}L_i\left(\log l_N(\theta)\right).$$

The necessary statistic provides again condensed information about the "shape" of likelihood $\log l_N(\theta)$ though this information need not be sufficient for its complete restoration.

Likelihood Matching. We apply the linear functionals $L_i(\cdot)$, $i = 1,\ldots,n$ to both sides of the Pythagorean relationship (3.81)

$$K(r_N{:}s_\theta) = K(r_N{:}s_{\theta,\hat\lambda}) + D(s_{\theta,\hat\lambda}\|s_\theta).$$

Substituting from (2.33) and (4.16)

$$L_i\left(K(r_N{:}s_\theta)\right) = L_i\left(-\frac{1}{N}\log l_N(\theta)\right),$$

$$L_i\left(D(s_{\theta,\hat\lambda}\|s_\theta)\right) = L_i\left(-\frac{1}{N}\log \hat l_N(\theta)\right),$$

we have

$$L_i\left(-\frac{1}{N}\log l_N(\theta)\right) = L_i\left(K(r_N{:}s_{\theta,\hat\lambda})\right) + L_i\left(-\frac{1}{N}\log \hat l_N(\theta)\right). \tag{4.60}$$

It can be shown quite analogously as in the case of independent observations that $L_i\left(K(r_N{:}s_{\theta,\hat\lambda})\right) = 0$. Then it follows directly from the last formula that

$$\boxed{L_i\left(\log l_N(\theta)\right) = L_i\left(\log \hat l_N(\theta)\right)} \tag{4.61}$$

for $i = 1,\ldots,n$. Thus, the use of a necessary statistic ensures that a perfect matching between the true and approximate log-likelihoods is attained at least locally—through (4.61).

4.6 Examples: Models and Statistics

Example 4.3 (*Various noise distributions*) Consider a sequence of independent and identically distributed data $y_1,\ldots,y_N$ with the following density functions

(a) normal distribution

$$s_{\mu,\lambda}(y) = \frac{1}{\sqrt{2\pi}\lambda}\exp\left(-\frac{1}{2\lambda^2}(y-\mu)^2\right);$$

(b) Cauchy distribution

$$s_{\mu,\lambda}(y) = \frac{1}{\pi\lambda}\frac{\lambda^2}{\lambda^2+(y-\mu)^2};$$

(c) double exponential distribution

$$s_{\mu,\lambda}(y) = \frac{1}{2\lambda} \exp\left(-\frac{1}{\lambda}|y-\theta|\right).$$

Taking the first-order derivatives of the above log-densities with respect to the *location parameter* μ at selected points $\mu_1^*, \ldots, \mu_n^*$ given a certain value λ^* of the scale parameter λ

$$h_i(y) = \frac{\partial}{\partial \mu} \log s_{\mu,\lambda}(y)\Big|_{\mu=\mu_i^*, \lambda=\lambda^*}$$

we obtain the following statistics

(a) normal distribution

$$h_i(y) = \frac{1}{(\lambda^*)^2}(y - \mu_i^*);$$

(b) Cauchy distribution

$$h_i(y) = \frac{2(y - \mu_i^*)}{(\lambda^*)^2 + (y - \mu_i^*)^2};$$

(c) double exponential distribution

$$h_i(y) = \frac{1}{\lambda^*} \text{sign}(y - \mu_i^*)$$

where sign($\cdot$) stands for the signum function, i.e., sign(x) = 1 for $x > 0$, sign(x) = -1 for $x < 0$ and sign(x) = 0 for $x = 0$ (the derivative of $|y - \mu|$ is not defined at $\theta = y$ but as the point has a Lebesgue measure 0, the value of the derivative can be set basically arbitrarily).

The functions are plotted in Fig. 4.15 for $\mu_1^* = 0$ (solid lines), $\mu_2^* = -2$, $\mu_3^* = +2$ (dashed lines) given $\lambda^* = 1$.

For the normal distribution the functions $h_i(y)$ are linearly dependent. Thus, it is sufficient to use only one of the functions.

Note that the functions $h_i(y)$ are

(a) linear for normal distribution,

(b) decreasing to zero given large values of y for Cauchy distribution,

(c) taking on basically two extreme values for double exponential distribution.

An occasional huge value of y thus affects the value of $\bar{h}_N$

(a) significantly for normal distribution,

(b) negligibly for Cauchy distribution,

(c) with limited effect for double exponential distribution.

Not surprisingly, the score functions $h_i(y)$ play an important role in the theory of robust statistics (Huber, 1981; Hampel et al., 1986).

Taking the first-order derivatives of the above log-densities with respect to the *scale parameter* λ at selected points $\lambda_1^*, \ldots, \lambda_n^*$ given a certain value μ^* of the location parameter μ

$$h_i(y) = \frac{\partial}{\partial \lambda} \log s_{\mu,\lambda}(y)\Big|_{\mu=\mu^*, \lambda=\lambda_i^*}$$

we obtain the following statistics

(a) normal distribution

$$h_i(y) = -\frac{1}{\lambda_i^*}\left(1 - \frac{(y-\mu^*)^2}{(\lambda_i^*)^2}\right);$$

(b) Cauchy distribution

$$h_i(y) = -\frac{1}{\lambda_i^*}\left(1 - \frac{2(y-\mu^*)^2}{(\lambda_i^*)^2 + (y-\mu^*)^2}\right);$$

(c) double exponential distribution

$$h_i(y) = -\frac{1}{\lambda_i^*}\left(1 - \frac{|y-\mu^*|}{\lambda_i^*}\right).$$

The functions are plotted in Fig. 4.16 for $\lambda_1^* = 1$ (solid lines), $\lambda_2^* = 0.8$, $\lambda_3^* = 1.2$ (dashed lines) given $\mu^* = 0$.

Note that for the normal distribution, the functions $h_i(y)$ are linearly dependent again; we can pick any one of these.   □

Example 4.4 (*Cauchy distribution*) Consider a sequence of independent Cauchy-distributed random variables $Y_k \sim C(\theta, 1)$ with a common density function

$$s_\theta(y) = \frac{1}{\pi}\frac{1}{1+(y-\theta)^2}.$$

We illustrate how different constructions of a single-data statistic $h_i(y)$ affect its shape. Differentiating the log-density $\log s_\theta(y)$ with respect to θ, we obtain

$$h_i(y) = \frac{d}{d\theta}\log s_\theta(y)\Big|_{\theta=\theta_i^*}$$
$$= \frac{2(y-\theta_i^*)}{1+(y-\theta_i^*)^2}.$$

An example of the statistic $h_1(y)$ for $\theta_1^* = 0$ is shown in the first plot in Fig. 4.17.

Taking the difference of log-densities $\log s_\theta(y)$ for two different values of θ, we get

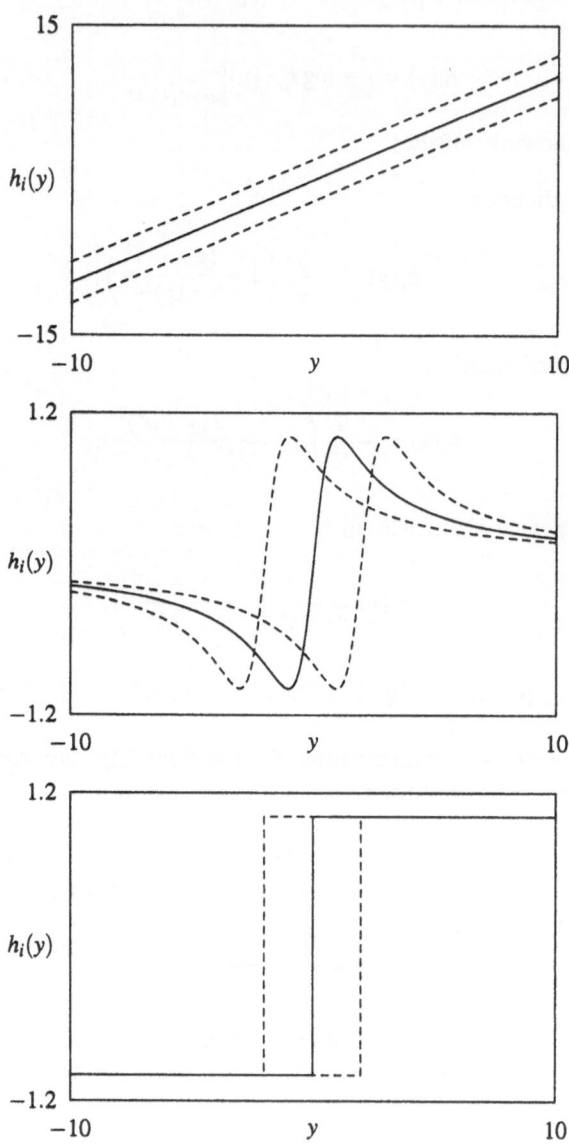

Fig. 4.15. Single-data statistics $h_i(y)$ for estimation of the location parameter of normal, Cauchy and double exponential distributions.

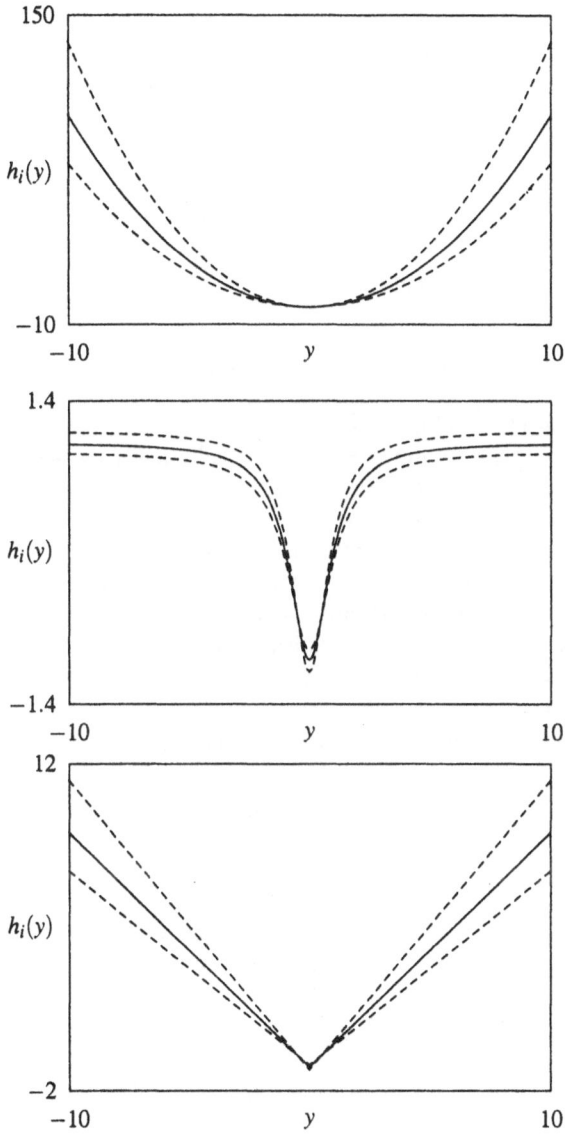

Fig. 4.16. Single-data statistics $h_i(y)$ for estimation of the scale parameter of normal, Cauchy and double exponential distributions.

$$h_i(y) = \log s_{\theta^*_{i+1}}(y) - \log s_{\theta^*_i}(y)$$

$$= \log \frac{1 + (y - \theta^*_i)^2}{1 + (y - \theta^*_{i+1})^2}.$$

An example of the statistic $h_1(y)$ for $\theta^*_1 = -3$, $\theta^*_2 = +3$ is shown in the second plot in Fig. 4.17.

Integrating the log-density $\log s_\theta(y)$ with a weighting function $w_i(\theta)$, we have

$$h_i(y) = \int w_i(\theta) \log s_\theta(y) \, d\theta$$

$$= -\int w_i(\theta) \log\left(1 + (y - \theta)^2\right) d\theta.$$

An example of the statistic $h_1(y)$ for

$$w^*_1(\theta) = \begin{cases} -1 & \text{if} \quad \theta \in [-5, 0) \\ 1 & \text{if} \quad \theta \in [0, 5] \end{cases}$$

is shown in the third plot in Fig. 4.17.

Note that the "shape" of the three statistics is basically the same. The difference between taking derivatives and differences is significant only when the "grid" points are far from each other. Similarly, the use of weighted integration provides statistics significantly different from those obtained by differentiating only when the weighting function is "wide" enough. Note that the statistics produced through weighted integration are typically flatter than those obtained by differentiation. This may be advantageous if the dimension n of the statistic $h(y)$ is relatively small (remember the functions $h_i(y)$ become a canonical statistic of the approximating exponential family).

For comparison, when calculating the second-order derivative of the log-density $\log s_\theta(y)$ with respect to θ, we obtain the following function

$$h_i(y) = \frac{d^2}{d\theta^2} \log s_\theta(y)\Big|_{\theta = \theta^*_i}$$

$$= -2 \frac{1 - (y - \theta^*_1)^2}{\left(1 + (y - \theta^*_1)^2\right)}.$$

An example of the statistic $h_1(y)$ for $\theta^*_1 = 0$ is shown in Fig. 4.18.   □

Example 4.5 (*Mixture of normal densities*) Consider a mixture family composed of densities in the form

$$s_\theta(y) = \theta s_1(y) + (1 - \theta) s_2(y)$$

where the densities $s_1(y)$ and $s_2(y)$ are normal

$$s_i(y) = \frac{1}{\sqrt{2\pi}} \exp\left(-\frac{1}{2}(y - \mu_i)^2\right)$$

with $\mu_1 = 1$ and $\mu_2 = -1$.

A set of single-data statistics $h_i(y)$ constructed by differentiating the log-density $\log s_\theta(y)$ with respect to θ

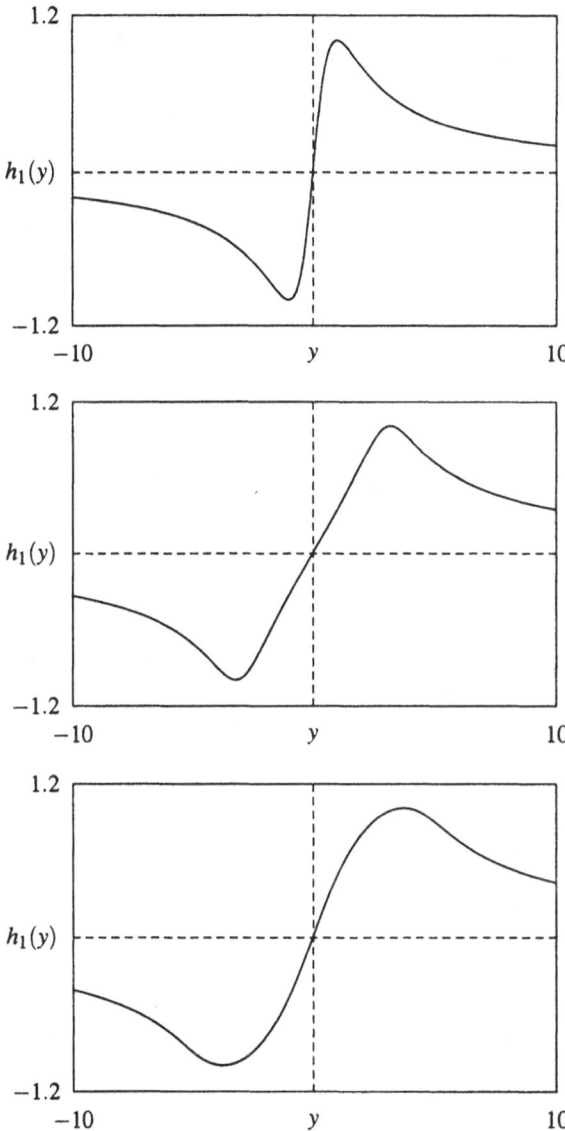

Fig. 4.17. Cauchyian noise: different single-data statistics $h_1(y)$ obtained by taking derivatives, differences and weighted integrals of the log-density $\log s_\theta(y)$ with respect to θ, respectively. For the sake of easier comparison, the statistics are normalized to have maximum equal to 1.

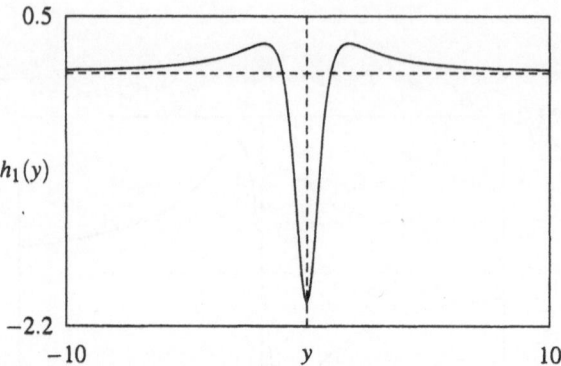

Fig. 4.18. Cauchyian noise: a single-data statistic $h_1(y)$ obtained by second-order differentiation of the log-density $\log s_\theta(y)$ with respect to θ.

$$h_i(y) = \frac{d}{d\theta} \log s_\theta(y) \bigg|_{\theta = \theta_i^*} = \frac{s_1(y) - s_2(y)}{\theta_i^* s_1(y) + (1 - \theta_i^*) s_2(y)}.$$

is shown in Fig. 4.19 for $\theta_1^* = 0.3$, $\theta_2^* = 0.5$, $\theta_3^* = 0.7$.   □

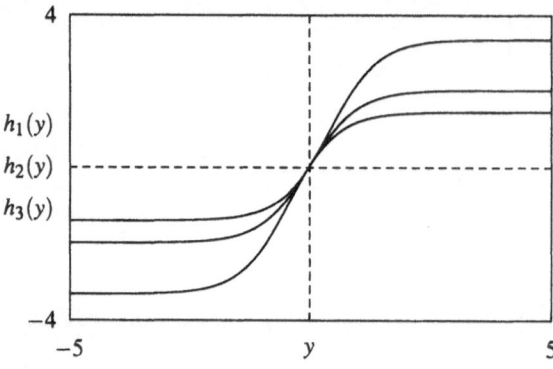

Fig. 4.19. Mixture of normal densities: single data statistics $h_i(y)$ for estimation of the weight in a mixture. The functions do not cross but they are tangent to each other at the point $(0,0)$.

Example 4.6 (*Nonlinear regression*) Consider a class of models

$$Y_k = f(Z_k; \theta) + E_k, \quad E_k \sim N(0, \sigma^2) \tag{4.62}$$

where $f(\cdot)$ is a known "response" function. For simplicity, f is supposed scalar.

The special instances of the model (4.62) is model linear in both the parameters and the regressor

$$f(z;\theta) = \theta^T z$$

and model linear in the parameters only

$$f(z;\theta) = \theta^T g(z).$$

Here we are interested in a more general case when $f(z;\theta)$ is nonlinear in both arguments but smooth enough in θ. This covers a large class of models of practical interest, including, e.g., a feedforward neural network with radial basis functions

$$f(z;\theta) = \sum_{i=1}^{M} w_i F(\|z - c_i\|), \quad \theta = (w_1,\ldots,w_M,c_1,\ldots,c_M)$$

where the unknown parameters are composed of the weights $w_1,\ldots,w_M \in \mathbb{R}$ and the centres $c_1,\ldots,c_M \in \mathbb{R}^{\dim z}$. A variation of the last model is a network with "ridge" functions

$$f(z;\theta) = \sum_{i=1}^{M} w_i F(\|v_i^T z\|), \quad \theta = (w_1,\ldots,w_M,v_1,\ldots,v_M)$$

where the unknown parameters are the weights $w_1,\ldots,w_M \in \mathbb{R}$ and the normal vectors $v_1,\ldots,v_M \in \mathbb{R}^{\dim z}$ of the ridges. In both cases, $F(\cdot)$ stands for some fixed smooth nonlinearity.

The model (4.62) induces the conditional theoretical density in the form

$$s_\theta(y|z) = \frac{1}{\sqrt{2\pi\sigma^2}} \exp\left(-\frac{1}{2\sigma^2}\left(y - f(z;\theta)\right)^2\right).$$

Given this model, the single-data statistics constructed by differencing of the log-density $\log s_\theta(y|z)$ take the following general form

$$
\begin{aligned}
h_i(y,z) &= \log s_{\theta_{i+1}^*}(y|z) - \log s_{\theta_i^*}(y|z) \\
&= \frac{1}{2\sigma^2}\left(\left(y - f(z;\theta_i^*)\right)^2 - \left(y - f(z;\theta_{i+1}^*)\right)^2\right).
\end{aligned}
$$

It is easy to see that the statistic of the whole sample is then equivalent to evaluating the mean square residuals

$$\frac{1}{N}\sum_{k=m+1}^{N+m} \left(y_k - f(z_k;\theta_i^*)\right)^2$$

for $n+1$ *fixed* models with parameters $\theta_1^*, \theta_2^*, \ldots, \theta_{n+1}^*$.

For comparison, if the single-data statistics are constructed by differentiating the log-density $\log s_\theta(y|z)$, they take the following form

$$
\begin{aligned}
h_i(y,z) &= \omega_i^{*T}\nabla_\theta \log s_{\theta_i^*}(y|z) \\
&= \frac{1}{\sigma^2}\left(y - f(z;\theta_i^*)\right)\left(\omega_i^{*T}\nabla_\theta f(z;\theta_i^*)\right).
\end{aligned}
$$

The full-sample statistic is then equivalent to evaluating directional derivatives of the mean square residuals

$$\omega_i^{*T} \nabla_\theta \left(\frac{1}{N} \sum_{k=m+1}^{N+m} (y_k - f(z_k; \theta_i^*))^2 \right)$$

at n fixed points $\theta_1^*, \ldots, \theta_n^*$ and for n fixed directions $\omega_1^*, \ldots, \omega_n^*$.

Analogous connections can be shown to hold for the single-data statistics obtained by higher-order differentiation or weighted integration. □

Example 4.7 (*Fixed nonlinearity*) Consider the model

$$Y_k = \text{atan}(\theta Z_k) + E_k, \quad E_k \sim N(0,1)$$

where atan($\cdot$) stands for the arctangent (inverse tangent) as an example of smooth nonlinearity. The model implies the conditional theoretical density in the form

$$s_\theta(y|z) = \frac{1}{\sqrt{2\pi}} \exp\left(-\frac{1}{2}(y - \text{atan}(\theta z))^2\right).$$

A set of single-data statistics $h_i(y,z)$ constructed by differentiating the log-density $\log s_\theta(y|z)$ with respect to θ

$$h_i(y,z) = \frac{d}{d\theta} \log s_\theta(y|z) \Big|_{\theta=\theta_i^*} = (y - \text{atan}(\theta_i^* z)) \frac{z}{1 + \theta_i^* z^2}.$$

is shown in Fig. 4.20 for $\theta_1^* = 0.5$, $\theta_2^* = 1$, $\theta_3^* = 2$. □

Example 4.8 (*Periodic signal*) Consider the model

$$Y_k = \sin(\theta k) + E_k, \quad E_k \sim N(0,1)$$

where sin($\cdot$) stands for the sinus as an example of periodic signal with unknown frequency. The model implies the conditional theoretical density in the form

$$s_\theta(y|z) = \frac{1}{\sqrt{2\pi}} \exp\left(-\frac{1}{2}(y - \sin(\theta z))^2\right)$$

with $z_k = k$.

A set of single-data statistics $h_i(y,z)$ constructed by differentiating the log-density $\log s_\theta(y|z)$ with respect to θ

$$h_i(y,z) = \frac{d}{d\theta} \log s_\theta(y|z) \Big|_{\theta=\theta_i^*} = (y - \sin(\theta_i^* z)) \cos(\theta_i^* z) z.$$

is shown in Fig. 4.21 for $\theta_1^* = 0.5$, $\theta_2^* = 1$, $\theta_3^* = 2$.

Note that due to a very special definition of $z_k = k$, this is one of rare occasions when the marginal density $\bar{r}_N(z)$ is known. Clearly, it holds

$$E_N(f(Z)) = \frac{1}{N} \sum_{k=m+1}^{N+m} f(z_k)$$

for any function $f(z)$. Thus, in this particular case we can use for estimation directly the formula (3.54). □

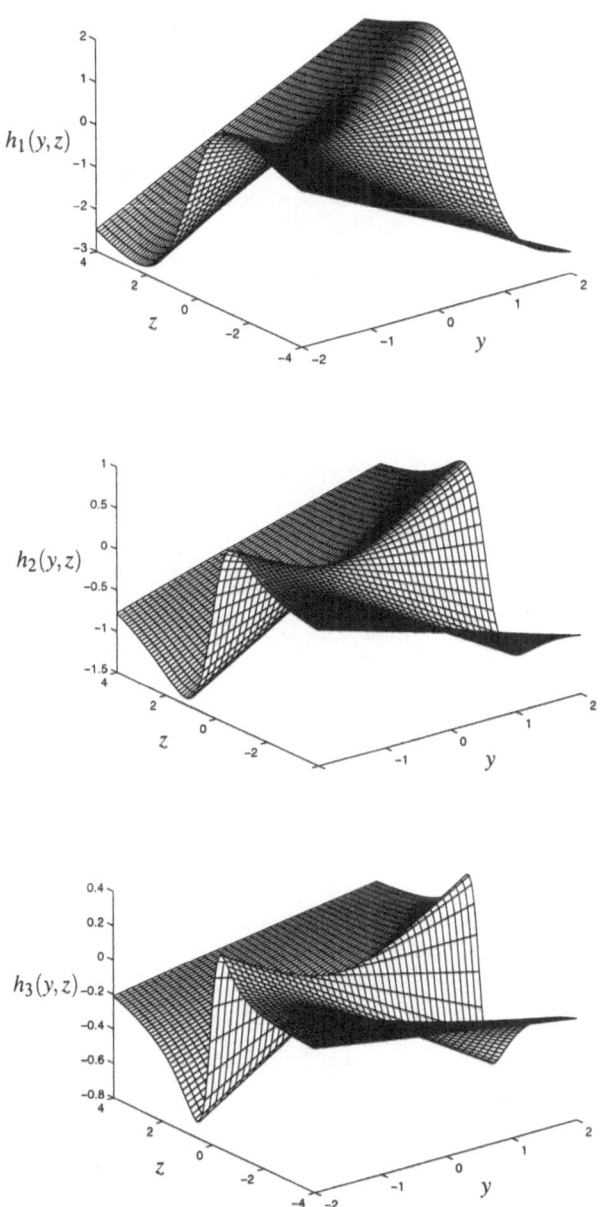

Fig. 4.20. Fixed nonlinearity: single-data statistics $h_i(y,z)$ for estimation of the nonlinear regression coefficient.

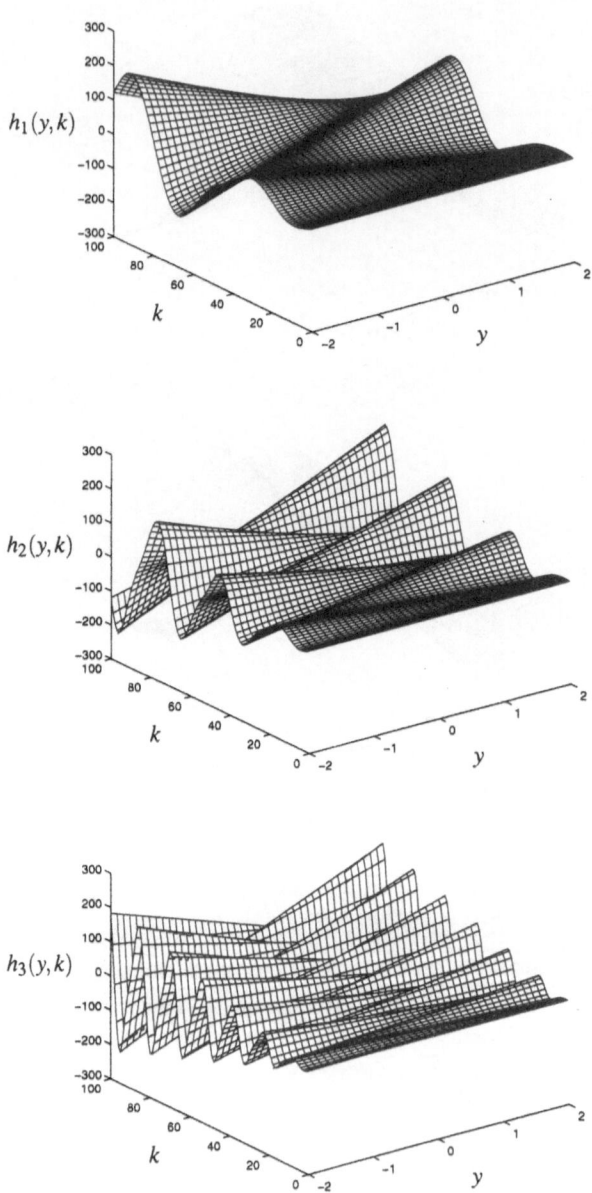

Fig.4.21. Periodic signal: single-data statistics $h_i(y,k)$ for estimation of the signal frequency.

4.7 Approximation of Empirical Expectation

We have seen that a kind of the Pythagorean relationship can be obtained even if the 'reference' function $w_{\theta,\tau}(z)$ or $s_\theta(y|z)$ is not normalized with respect to z or (y,z), respectively. As soon as we accept this extension , it is tempting to use the methodology developed in Sect. 3.3 for approximate empirical expectation of *any* distance or loss function that depends at the same time on data and some unknown or free parameter.

A typical example is when we are to make an optimum decision a so to minimize the empirical expectation of a certain loss function $L(y;a)$

$$\frac{1}{N} \sum_{k=1}^{N} L(y_k;a)$$

given the value of a data statistic $T_N(y^N)$ rather than the complete sample y^N. The approximation shown below might thus appear useful in empirical decision-making (Robbins, 1956; Robbins, 1964) or in calculation of robust M-estimates (Huber, 1981; Hampel et al., 1986).

Note that we deliberately use below the same notation as in Sect. 3.3 as the deductions go formally along the same lines and it is easier for the reader to follow the analogies.

Independent Observations

Consider the problem of computing the empirical expectation

$$E_N\big(L(Y;\theta)\big) = \int r_N(y) L(y;\theta) \, dy$$

$$= \frac{1}{N} \sum_{k=1}^{N} L(y_k;\theta)$$

of a certain function $L(y;\theta)$.

Inaccuracy. We introduce the following function

$$s_\theta(y) = \exp\big(-L(y;\theta)\big).$$

The function is clearly nonnegative but, as a rule, not normalized so it cannot be regarded, even formally, as a density function.

Yet, extending the concept of inaccuracy by admitting the second argument to be unnormalized

$$K(r_N:s_\theta) = \int r_N(y) \log \frac{1}{s_\theta(y)} \, dy,$$

we can still write

$$E_N\big(L(Y;\theta)\big) = K(r_N:s_\theta).$$

In a way completely analogous to Sect. 3.3, we prove that an analogue of the Pythagorean relationship holds even in this case.

Exponential Family. First, we define an exponential family $\mathcal{S}_{\theta;h}$ composed of density functions

$$s_{\theta,\lambda}(y) = s_\theta(y) \exp\left(\lambda^T h(y) - \psi(\theta,\lambda)\right)$$

where $h(y)$ is a properly chosen vector function of y (see below) and

$$\psi(\theta,\lambda) = \log \int s_\theta(y) \exp\left(\lambda^T h(y)\right) dy$$

is logarithm of the normalizing divisor.

Pythagorean Relationship. Suppose there exists a h-projection $s_{\theta,\hat\lambda}(y)$ of the empirical density $r_N(y)$ onto the exponential family $\mathcal{S}_{\theta;h}$ such that

$$\int s_{\theta,\hat\lambda}(y)\, h(y)\, dy = \int r_N(y)\, h(y)\, dy.$$

Then we can write

$$
\begin{aligned}
& K(r_N{:}s_\theta) - K(r_N{:}s_{\theta,\hat\lambda}) \\
&= \int r_N(y) \log \frac{s_{\theta,\hat\lambda}(y)}{s_\theta(y)}\, dy \\
&= \int r_N(y) \log \frac{s_\theta(y) \exp\left(\hat\lambda^T h(y) - \psi(\theta,\hat\lambda)\right)}{s_\theta(y)}\, dy \\
&= \hat\lambda^T \left(\int r_N(y)\, h(y)\, dy \right) - \psi(\theta,\hat\lambda) \\
&= \hat\lambda^T \left(\int s_{\theta,\hat\lambda}(y)\, h(y)\, dy \right) - \psi(\theta,\hat\lambda) \\
&= \int s_{\theta,\hat\lambda}(y) \log \frac{s_\theta(y) \exp\left(\hat\lambda^T h(y) - \psi(\theta,\hat\lambda)\right)}{s_\theta(y)}\, dy \\
&= \int s_{\theta,\hat\lambda}(y) \log \frac{s_{\theta,\hat\lambda}(y)}{s_\theta(y)}\, dy \\
&= D(s_{\theta,\hat\lambda} \| s_\theta).
\end{aligned}
$$

As a result, we obtain the identity

$$\boxed{K(r_N{:}s_\theta) = K(r_N{:}s_{\theta,\hat\lambda}) + D(s_{\theta,\hat\lambda} \| s_\theta)} \tag{4.63}$$

which can be regarded as yet another analogue of the Pythagorean relationship for the case of an unnormalized $s_\theta(y)$.

Minimum K-L Distance. Owing to the following identity

$$0 = \min_{\lambda} D(s_{\theta,\hat{\lambda}} \| s_{\theta,\lambda})$$

$$= \min_{\lambda} \int s_{\theta,\hat{\lambda}}(y) \log \frac{s_{\theta,\hat{\lambda}}(y)}{s_{\theta}(y) \exp\left(\lambda^T h(y,z) - \psi(\theta,\lambda)\right)} \, dy$$

$$= D(s_{\theta,\hat{\lambda}} \| s_{\theta}) - \max_{\lambda} \left(\lambda^T \hat{h}(\theta,\hat{\lambda}) - \psi(\theta,\lambda)\right)$$

$$= D(s_{\theta,\hat{\lambda}} \| s_{\theta}) - \max_{\lambda} \left(\lambda^T \bar{h}_N - \psi(\theta,\lambda)\right),$$

we can compute the minimum K-L distance by maximizing $\lambda^T \bar{h}_N - \psi(\theta,\lambda)$ over λ

$$D(s_{\theta,\hat{\lambda}} \| s_{\theta}) = \max_{\lambda} \left(\lambda^T \bar{h}_N - \psi(\theta,\lambda)\right). \qquad (4.64)$$

Choice of Statistic. A proper choice of the function $h(y)$ can make the inaccuracy $K(r_N : s_{\theta,\hat{\lambda}})$ in (4.63) fully or almost independent of the unknown parameter θ. In such a case, we can use the approximation

$$E_N\left(L(Y;\theta)\right) \approx C + \max_{\lambda} \left(\lambda^T \bar{h}_N - \psi(\theta,\lambda)\right) \qquad (4.65)$$

where C is a constant independent of θ.

Using arguments analogous to those in Sect. 3.5 and 4.5, we find that it makes sense to choose $h_i(y)$, $i = 1,\dots,n$ as linearly independent functions from the linear space spanned by the functions $L(y;\theta) - L(y;\theta_0)$ for all $\theta \in \mathcal{T}$ where θ_0 is an arbitrary fixed point in $\mathcal{T}$.

Such functions can constructed, for instance, by taking differences, derivatives or weighted integrals of $L(y;\theta)$

$$h_i(y) = L(y;\theta_{i+1}^*) - L(y;\theta_i^*)$$
$$h_i(y) = \omega_i^{*T} \nabla_\theta L(y;\theta_i^*)$$
$$h_i(y) = \int w_i^*(\theta) L(y;\theta) \, d\theta,$$

see (4.48), (4.49), (4.50) for more details.

Example 4.9 (*Quadratic function*) Consider the function

$$L(y;\theta) = \frac{1}{2}(y-\theta)^2.$$

The maximization (4.64) with $h(y) = y$ gives

$$\max_{\lambda} \left(\lambda E_N(Y) - \log \int \exp\left(-\frac{1}{2}(y-\theta)^2\right) \exp(\lambda y) \, dy\right)$$

$$= \max_{\lambda} \left(\lambda E_N(Y) - \log \int \exp\left(-\frac{1}{2}(y-(\theta+\lambda))^2\right) dy - \frac{1}{2}(\theta^2 - (\theta+\lambda)^2)\right)$$

$$= \max_{\lambda} \left(\lambda E_N(Y) - \frac{1}{2}\log 2\pi - \frac{1}{2}\lambda^2 - \theta\lambda \right)$$

$$= -\frac{1}{2}\log 2\pi + \frac{1}{2}\left(\theta - E_N(Y) \right)^2.$$

The result coincides, up to an additive constant independent of θ, with the true expectation $E_N\left(L(Y;\theta)\right)$.   □

Controlled Dynamic Systems

Consider the problem of computing the empirical expectation

$$E_N\left(L(Y,Z;\theta)\right) = \iint r_N(y,z)\,L(y,z;\theta)\,dy\,dz$$

$$= \frac{1}{N}\sum_{k=m+1}^{N+m} L(y_k,z_k;\theta)$$

of a certain function $L(y,z;\theta)$.

Inaccuracy. Once again we start by introducing the function

$$s_\theta(y,z) = \exp\left(-L(y,z;\theta)\right)$$

which is nonnegative but unnormalized in general.

Defining inaccuracy relative to $s_\theta(y,z)$ as

$$K(r_N:s_\theta) = \iint r_N(y,z)\,\log\frac{1}{s_\theta(y,z)}\,dy\,dz,$$

we can write

$$E_N\left(L(Y,Z;\theta)\right) = K(r_N:s_\theta).$$

Exponential Family. Next we define an exponential family $S_{\theta;h}$ composed of density functions

$$s_{\theta,\lambda}(y,z) = s_\theta(y,z)\,\exp\left(\lambda^T h(y,z) - \psi(\theta,\lambda)\right)$$

where $h(y,z)$ is a properly chosen vector function of (y,z) (see below) and

$$\psi(\theta,\lambda) = \log \iint s_\theta(y,z)\,\exp\left(\lambda^T h(y,z)\right)\,dy\,dz$$

is logarithm of the normalizing divisor.

Pythagorean Relationship. Suppose there exists a h-projection $s_{\theta,\hat{\lambda}}(y,z)$ of the empirical density $r_N(y,z)$ onto the exponential family $S_{\theta;h}$ that satisfies

$$\iint s_{\theta,\hat{\lambda}}(y,z)\,h(y,z)\,dy\,dz = \iint r_N(y,z)\,h(y,z)\,dy\,dz.$$

The above implies

$$K(r_N{:}s_\theta) - K(r_N{:}s_{\theta,\hat\lambda})$$

$$= \iint r_N(y,z) \log \frac{s_{\theta,\hat\lambda}(y,z)}{s_\theta(y,z)}\, dy\, dz$$

$$= \iint r_N(y,z) \log \frac{s_\theta(y,z) \exp(\hat\lambda^T h(y,z) - \psi(\theta,\hat\lambda))}{s_\theta(y,z)}\, dy\, dz$$

$$= \hat\lambda^T \left(\iint r_N(y,z) h(y,z)\, dy\, dz \right) - \psi(\theta,\hat\lambda)$$

$$= \hat\lambda^T \left(\iint s_{\theta,\hat\lambda}(y,z) h(y,z)\, dy\, dz \right) - \psi(\theta,\hat\lambda)$$

$$= \iint s_{\theta,\hat\lambda}(y,z) \log \frac{s_\theta(y,z) \exp(\hat\lambda^T h(y,z) - \psi(\theta,\hat\lambda))}{s_\theta(y,z)}\, dy\, dz$$

$$= \iint s_{\theta,\hat\lambda}(y,z) \log \frac{s_{\theta,\hat\lambda}(y,z)}{s_\theta(y,z)}\, dy\, dz$$

$$= D(s_{\theta,\hat\lambda} \| s_\theta).$$

Hence, we have another analogue of the Pythagorean relationship for the case of an unnormalized $s_\theta(y,z)$

$$\boxed{K(r_N{:}s_\theta) = K(r_N{:}s_{\theta,\hat\lambda}) + D(s_{\theta,\hat\lambda} \| s_\theta).} \tag{4.66}$$

Minimum K-L Distance. The obvious identity

$$0 = \min_\lambda D(s_{\theta,\hat\lambda} \| s_{\theta,\lambda})$$

$$= \min_\lambda \iint s_{\theta,\hat\lambda}(y,z) \log \frac{s_{\theta,\hat\lambda}(y,z)}{s_\theta(y,z) \exp(\lambda^T h(y,z) - \psi(\theta,\lambda))}\, dy\, dz$$

$$= D(s_{\theta,\hat\lambda} \| s_\theta) - \max_\lambda \left(\lambda^T \hat h(\theta,\hat\lambda) - \psi(\theta,\lambda) \right)$$

$$= D(s_{\theta,\hat\lambda} \| s_\theta) - \max_\lambda \left(\lambda^T \bar h_N - \psi(\theta,\lambda) \right)$$

implies formally the same result as above

$$\boxed{D(s_{\theta,\hat\lambda} \| s_\theta) = \max_\lambda \left(\lambda^T \bar h_N - \psi(\theta,\lambda) \right).} \tag{4.67}$$

Choice of Statistic. When the function $h(y,z)$ is chosen so that the inaccuracy $K(r_N{:}s_{\theta,\hat\lambda})$ in (4.66) is almost independent of the unknown parameter θ, we can use the approximation

$$\boxed{E_N\left(L(Y,Z;\theta) \right) \approx C + \max_\lambda \left(\lambda^T \bar h_N - \psi(\theta,\lambda) \right).} \tag{4.68}$$

where C is a constant independent of θ.

Similarly as in Sect. 3.5 and 4.5, we find that $h_i(y,z)$, $i = 1,\ldots,n$ should be chosen as linearly independent functions from the linear space spanned by the functions $L(y,z;\theta) - L(y,z;\theta_0)$ for all $\theta \in \mathcal{T}$ where θ_0 is an arbitrary fixed point in $\mathcal{T}$.

Such function can be obtained, for instance, by taking differences, derivatives or weighted integrals of $L(y,z;\theta)$

$$h_i(y,z) = L(y,z;\theta_{i+1}^*) - L(y,z;\theta_i^*)$$
$$h_i(y,z) = \omega_i^{*T} \nabla_\theta L(y,z;\theta_i^*)$$
$$h_i(y,z) = \int w_i^*(\theta) L(y,z;\theta) \, d\theta,$$

see (4.56), (4.57), (4.58) for more details.

Example 4.10 (*Quadratic form*) Suppose the function

$$L(y,z;\theta) = \frac{1}{2}(y - \theta z)^2.$$

The maximization (4.67) with $h_1(y,z) = yz$ and $h_2(y,z) = z^2$ gives

$$\max_\lambda \left(\lambda_1 E_N(YZ) + \lambda_2 E_N(Z^2) \right.$$
$$\left. - \log \iint \exp\left(-\frac{1}{2}(y - \theta z)^2\right) \exp(\lambda_1 yz + \lambda_2 z^2) \, dy \, dz \right)$$
$$= \max_\lambda \left(\lambda_1 E_N(YZ) + \lambda_2 E_N(Z^2) \right.$$
$$\left. - \log \iint \exp\left(-\frac{1}{2} \begin{bmatrix} y \\ z \end{bmatrix}^T \begin{bmatrix} 1 & -\theta - \lambda_1 \\ -\theta - \lambda_1 & \theta^2 - 2\lambda_2 \end{bmatrix} \begin{bmatrix} y \\ z \end{bmatrix}\right) dy \, dz \right)$$
$$= \max_\lambda \left(\lambda_1 E_N(YZ) + \lambda_2 E_N(Z^2) - \log 2\pi \left| \begin{matrix} 1 & -\theta - \lambda_1 \\ -\theta - \lambda_1 & \theta^2 - 2\lambda_2 \end{matrix} \right|^{-\frac{1}{2}} \right)$$
$$= \max_\lambda \left(\lambda_1 E_N(YZ) + \lambda_2 E_N(Z^2) - \log 2\pi - \frac{1}{2} \log(-\lambda_1^2 - 2\theta\lambda_1 - 2\lambda_2) \right)$$
$$= -\frac{1}{2} - \log 2\pi - \frac{1}{2} \log E_N(Z^2) + \frac{1}{2} E_N(Z^2) \left(\theta - \frac{E_N(YZ)}{E_N(Z^2)} \right)^2.$$

The result coincides again, up to an additive constant independent of θ, with the true expectation $E_N(L(Y,Z;\theta))$. $\qquad\square$

4.8 Historical Notes

Principle of Maximum Entropy/Minimum K-L Distance. The principle of maximum entropy first appeared in Jaynes (1957). A very readable account of the underlying theory can be found in Jaynes (1979). The principle of minimum K-L distance was introduced in Kullback (1959). Shore and Johnson (1980) formulated a set of axioms and showed that both the principles follow from the axioms. van Campenhout and Cover (1981) and Csiszár (1984) deduced the principles of maximum entropy and minimum K-L distance as a consequence of properties of conditional probability.

Minimum K-L Distance Distribution. A solution to the minimum K-L distance problem was given first in Kullback (1959) using the method of Lagrange multipliers. Csiszár (1984) solved rigorously the problem in a general setting.

In the case of Markov conditioning, a solution to the maximum conditional entropy problem was given first in Spitzer (1972) and Justesen and Høholdt (1984). A solution to the minimum conditional K-L distance problem was shown in Csiszár et al. (1987).

Large Deviation Theory. Sanov (1957) proved the first large deviation theorem for the case of independent and identically distributed random variables. His work was clarified and extended by Hoeffding (1965). Proofs of these results were combinatorial, using Stirling's formula. Another proof, using the method of types developed in Csiszár and Körner (1981), was given in Csiszár et al. (1987) and Cover and Thomas (1991).

To prove similar results for Markov chains, Boza (1971) and Natarajan (1985) used the counting approach developed in Whittle (1955) and Billingsley (1961). A considerably more general result was proved in Csiszár et al. (1987) using the method of types.

Large deviation theory for general Markov processes was developed in Donsker and Varadhan (1975a; 1975b), Donsker and Varadhan (1976), Donsker and Varadhan (1983) using advanced mathematical tools.

A readable introduction into the large deviation techniques can be found in Bucklew (1990). Ellis (1985) gives a more rigorous treatment in the context of statistical mechanics.

Possible applications of the large deviation approximation in Bayesian estimation were shown in Kulhavý (1994a) and Kulhavý and Hrnčíř (1994).

Necessary Statistic. The concept of necessary statistic was introduced by Dynkin (1951). A good account of the underlying theory that follows essentially Dynkin's work can be found in Zacks (1971).

5. Numerical Implementation

The use of approximations (4.8) and (4.15) is accompanied with massive drop in computational complexity. First, the approximations use a statistic of finite, limited dimension. Second, the dimension of the optimization problem invoked is given by the dimension of data entering model at one time instant only. Compare it with the ideal solution which in general requires all data to be stored and processed.

Yet, to solve the optimization problems (4.7) or (4.14) for all or many values of the unknown parameter, we may still need a lot of computing power. In this chapter we give some recommendations concerning efficient numerical implementation of the estimation algorithm. We also show how the estimation schemes (4.8) and (4.15) can be modified so to become able of tracking slowly varying parameters.

5.1 Direct Optimization

The most straightforward way of implementing (4.7) and (4.14) is to solve directly the optimization problems involved.

Independent Observations

Convex Minimization Problem. Since most optimization packages provide algorithms for finding minimum of a given function, we rewrite (4.7) explicitly as minimization problem

$$\min_{\lambda} J(\theta, \lambda)$$

where the optimized function is

$$J(\theta, \lambda) = -\lambda^T \bar{h}^N + \log \int s_\theta(y) \exp\left(\lambda^T h(y)\right) dy.$$

We have seen in Sect. 3.3 that the set $\mathcal{N}_\theta$ of all values of λ for which logarithm of the normalizing divisor is finite (3.26)

$$\psi(\theta, \lambda) = \log \int s_\theta(y) \exp\left(\lambda^T h(y)\right) dy < \infty$$

is convex. The function $\psi(\theta, \lambda)$ is a convex function of λ on $\mathcal{N}_\theta$. Since $\lambda^T \bar{h}^N$ is linear in λ, the function $J(\theta, \lambda)$ is convex as well. Hence, the problem we solve for every $\theta \in \mathcal{T}$ is to find the minimum of a convex function over a convex set.

Gradient and Hessian. The optimization problem shown above can be solved efficiently using gradient, quasi-Newton or Newton search methods. Incidentally, the results presented in this book were all obtained using the Broyden-Fletcher-Goldfarb-Shanno (BFGS) method.

The application of gradient and Newton methods is facilitated by the fact that it is easy, conceptually at least, to compute the first- and second-order partial derivatives of $J(\theta, \lambda)$ with respect to λ. In particular, the gradient of the function $J(\theta, \lambda)$ with respect to λ is

$$
\begin{aligned}
\nabla_\lambda J(\theta, \lambda) &= -\bar{h}_N + \int \frac{s_\theta(y) \exp(\lambda^T h(y)) \, dy}{\int s_\theta(y) \exp(\lambda^T h(y)) \, dy} h(y) \, dy \\
&= -\bar{h}_N + \hat{h}(\theta, \lambda).
\end{aligned}
$$

Thus, the gradient of $J(\theta, \lambda)$ is equal to the difference between the theoretical and empirical means of $h(Y)$

$$
\nabla_\lambda J(\theta, \lambda) = E_{\theta, \lambda}(h(Y)) - E_N(h(Y)). \tag{5.1}
$$

The Hessian, i.e., the matrix of the second-order derivatives of $J(\theta, \lambda)$ with respect to λ is given by

$$
\begin{aligned}
\nabla_\lambda^2 J(\theta, \lambda) &= \nabla_\lambda \hat{h}^T(\theta, \lambda) \\
&= \nabla_\lambda \int s_\theta(y) \exp(\lambda^T h(y) - \psi(\theta, \lambda)) \, h^T(y) \, dy \\
&= \int s_{\theta, \lambda}(y) \, (h(y) - \hat{h}(\theta, \lambda)) \, h^T(y) \, dy \\
&= \int s_{\theta, \lambda}(y) \, (h(y) - \hat{h}(\theta, \lambda)) \, (h(y) - \hat{h}(\theta, \lambda))^T \, dy \\
&= E_{\theta, \lambda}\left((h(Y) - \hat{h}(\theta, \lambda)) \, (h(Y) - \hat{h}(\theta, \lambda))^T \right).
\end{aligned}
$$

Hence, the Hessian of $J(\theta, \lambda)$ is the theoretical covariance of $h(Y)$

$$
\nabla_\lambda^2 J(\theta, \lambda) = \mathrm{Cov}_{\theta, \lambda}(h(Y)). \tag{5.2}
$$

Numerical Integration. The major source of computational complexity is the integration inherently involved in evaluating logarithm of the normalizing divisor $\psi(\theta, \lambda)$ and possibly in calculation of the mean $E_{\theta, \lambda}(h(Y))$ and covariance $\mathrm{Cov}_{\theta, \lambda}(h(Y))$ if needed.

When Y is scalar, standard quadrature formulae can be used. These methods fail, however, when the dimension of Y becomes larger than, say, 4–6. Then Monte Carlo techniques turn out to be more efficient.

The Monte Carlo computation of the normalizing divisor of the function $f_{\theta, \lambda}(y) = s_\theta(y) \exp(\lambda^T h(y))$ is based on the following approximation

$$
\int f_{\theta, \lambda}(y) \, dy = \int s(y) \frac{f_{\theta, \lambda}(y)}{s(y)} \, dy \approx \frac{1}{M} \sum_{k=1}^{M} \frac{f_{\theta, \lambda}(y_k)}{s(y_k)}
$$

where $s(y) > 0$ is a density function and $y_1,\ldots,y_M$ are independent samples drawn from $s(y)$. Of course, the choice of the density $s(y)$ is not arbitrary; the method is efficient only if $s(y)$ is close enough to $s_{\theta,\lambda}(y)$ (see, e.g., Ripley, 1987).

Provided $s_{\theta,\lambda}(y)$ is not too far from $s_{\theta}(y)$, we can also use the approximation

$$\int f_{\theta,\lambda}(y)\,dy \approx \frac{1}{M} \sum_{k=1}^{M} \exp(\lambda^T h(y_k))$$

where the samples $y_1,\ldots,y_M$ are distributed now according to $s_{\theta}(y)$.

Similarly, we can approximate the moments $E_{\theta,\lambda}(h(Y))$ and $\mathrm{Cov}_{\theta,\lambda}(h(Y))$.

Iterative Optimization. When the dimension of the statistic $h(y)$ is too large, it may be more efficient to organize the calculation of the minimum K-L distance $D(\mathcal{R}_N \| s_{\theta})$ so that we optimize *one entry* λ_i of the vector λ *at a time*

$$\min_{\lambda_i} \left(-\lambda^T \bar{h}^N + \log \int s_{\theta}(y) \, \exp(\lambda^T h(y))\,dy \right)$$

while the other entries λ_j, $j \neq i$ are fixed at their last values. The optimization is done for $i = 1,\ldots,n$ and then the whole loop is repeated—until the minimum is found with a prescribed precision.

The iterative procedure basically means that the constraints $E_{\theta,\hat{\lambda}}(h_i(Y)) = E_N(h_i(Y))$, $i = 1,\ldots,n$ are imposed one at a time. Owing to the convexity of K-L distance and linearity of the constraints, the whole procedure ultimately converges to the true solution (Csiszár, 1975).

Note that in the above scheme we need one integration to evaluate $J(\theta,\lambda)$, one integration to calculate $\frac{\partial}{\partial \lambda_i} J(\theta,\lambda)$ and one integration to calculate $\frac{\partial^2}{\partial \lambda_i^2} J(\theta,\lambda)$. On the other hand, when optimizing the whole vector λ at once, 1, n and $n(n+1)/2$ integrations are needed to calculate the functional value, gradient and Hessian, respectively. The minimum number of integrations per iteration in the former case is paid, however, by the increase of the number of iterations. Optimization in coordinate directions may run quite slowly.

The decision which method should be applied depends on a particular problem. Generally speaking, when the cost of integration is high, the iterative procedure turns out more efficient.

Controlled Dynamic Systems

Convex Minimization Problem. Once again, (4.14) can be rewritten as minimization problem

$$\min_{\lambda} J(\theta,\lambda)$$

with

$$J(\theta,\lambda) = -\lambda^T \bar{h}^N + \log \iint s_{\theta}(y|z) \, \exp(\lambda^T h(y,z))\,dy\,dz.$$

Again, as shown in Sect. 3.3, the set $\mathcal{N}_\theta$ of all values of λ for which logarithm of the normalizing divisor is finite (3.76)

$$\psi(\theta,\lambda) = \log \iint s_\theta(y) \exp\left(\lambda^T h(y,z)\right) dy\,dz < \infty$$

is convex. The function $\psi(\theta,\lambda)$ is a convex function of λ on $\mathcal{N}_\theta$. As $\lambda^T \bar{h}^N$ is linear in λ, the function $J(\theta,\lambda)$ is convex as well. Thus, the problem we solve for every $\theta \in \mathcal{T}$ is to find the minimum of a convex function over a convex set.

Gradient and Hessian. The application of gradient and Newton methods is again facilitated by the conceptually easy computation of the gradient and Hessian of $J(\theta,\lambda)$ with respect to λ.

In particular, the gradient of the function $J(\theta,\lambda)$ with respect to λ is

$$\begin{aligned}
\nabla_\lambda J(\theta,\lambda) &= -\bar{h}_N + \iint \frac{s_\theta(y|z) \exp\left(\lambda^T h(y,z)\right)}{\iint s_\theta(y|z) \exp\left(\lambda^T h(y,z)\right) dy\,dz} h(y,z)\,dy\,dz \\
&= -\bar{h}_N + \hat{h}(\theta,\lambda),
\end{aligned}$$

i.e., equal to the difference between the theoretical and empirical means of $h(Y,Z)$

$$\boxed{\nabla_\lambda J(\theta,\lambda) = E_{\theta,\lambda}\big(h(Y,Z)\big) - E_N\big(h(Y,Z)\big).} \tag{5.3}$$

The Hessian of $J(\theta,\lambda)$ with respect to λ is

$$\begin{aligned}
\nabla_\lambda^2 J(\theta,\lambda) &= \nabla_\lambda \hat{h}^T(\theta,\lambda) \\
&= \nabla_\lambda \iint s_\theta(y|z) \exp\left(\lambda^T h(y,z) - \psi(\theta,\lambda)\right) h^T(y,z)\,dy\,dz \\
&= \iint s_{\theta,\lambda}(y,z) \left(h(y,z) - \hat{h}(\theta,\lambda)\right) h^T(y,z)\,dy\,dz \\
&= \iint s_{\theta,\lambda}(y,z) \left(h(y,z) - \hat{h}(\theta,\lambda)\right) \left(h(y,z) - \hat{h}(\theta,\lambda)\right)^T dy\,dz \\
&= E_{\theta,\lambda}\left(\left(h(Y,Z) - \hat{h}(\theta,\lambda)\right)\left(h(y,z) - \hat{h}(\theta,\lambda)\right)^T\right),
\end{aligned}$$

i.e., equal to the theoretical covariance of $h(Y,Z)$

$$\boxed{\nabla_\lambda^2 J(\theta,\lambda) = \mathrm{Cov}_{\theta,\lambda}\big(h(Y,Z)\big).} \tag{5.4}$$

Numerical Integration. Integration involved in calculation of the function $\psi(\theta,\lambda)$ and possibly the mean $E_{\theta,\lambda}\big(h(Y,Z)\big)$ and covariance $\mathrm{Cov}_{\theta,\lambda}\big(h(Y,Z)\big)$ is taken over the space of all possible values of (Y,Z). Since the dimension of Z for practically interesting problems is usually beyond the margin when quadrature formulae can be used efficiently, Monte Carlo techniques can generally be recommended.

The Monte Carlo computation of the normalizing divisor of the function $f_{\theta,\lambda}(y,z) = s_\theta(y|z) \exp\left(\lambda^T h(y,z)\right)$ is based on the approximation

$$\iint f_{\theta,\lambda}(y)\,dy\,dz = \iint s(y,z)\,\frac{f_{\theta,\lambda}(y,z)}{s(y,z)}\,dy\,dz \approx \frac{1}{M}\sum_{k=1}^{M}\frac{f_{\theta,\lambda}(y_k,z_k)}{s(y_k,z_k)}$$

where $s(y,z) > 0$ is a density function and $(y_1,z_1),\ldots,(y_M,z_M)$ are independent samples drawn from $s(y,z)$. The density function $s(y,z)$ should be chosen close enough to $s_{\theta,\lambda}(y,z)$. In a similar way, we can approximate the moments $E_{\theta,\lambda}\big(h(Y,Z)\big)$ and $\mathrm{Cov}_{\theta,\lambda}\big(h(Y,Z)\big)$.

For more information about the advanced methods of multivariate integration in Bayesian statistics, see the references in Sect. 1.4.

Iterative Optimization. Owing to the higher dimension of (Y,Z), a considerably higher dimension n of the statistic $h(y,z)$ is typically needed as well. In such a case, the iterative procedure will generally be the preferred choice. The basic idea is again that we optimize *one entry* λ_i of the vector λ *at a time*

$$\min_{\lambda_i}\left(-\lambda^T \bar{h}^N + \log \iint s_\theta(y|z)\exp\big(\lambda^T h(y,z)\big)\,dy\,dz\right)$$

while the other entries λ_j, $j \neq i$ are fixed at the last values. Once again, owing to the convexity of K-L distance and linearity of the constraints $E_{\theta,\hat{\lambda}}\big(h_i(Y,Z)\big) = E_N\big(h_i(Y,Z)\big)$, the procedure is known to converge to the true solution.

5.2 Prior Approximation of Critical Maps

A natural way of avoiding onerous *on-line* computations is to precompute as much as possible beforehand. The approximate estimation schemes based on (4.7) and (4.14) are particularly well suited for this kind of implementation.

Independent Observations

Structure of Algorithm. The use of a fixed statistic $h(y)$ implies a fixed structure of the estimation algorithm as well. The core of approximate Bayesian estimation is in the following two steps which stand for data compression and approximate restoration of inaccuracy, respectively

$$y^N \mapsto \bar{h}_N \mapsto D(\mathcal{R}_N\|s_\theta).$$

The conceptual structure of the estimation algorithm is outlined in Fig. 5.1. Note that except a separate memory, all blocks in the algorithm are only functions of their input arguments; they do not contain any dynamic elements and can thus be analysed and possibly approximated before estimation actually starts.

(a) The transform $y_N \mapsto h(y_N)$ is usually simple but if the statistic $h(y)$ is defined through an integral transform of log-density $\log s_\theta(y)$, the step may call for numerical integration.

(b) The map $(\bar{h}_{N-1}, h(y_N)) \mapsto \bar{h}_N$ is just a trivial update of the statistic value.

(c) The map $\bar{h}_N \mapsto D(\mathcal{R}_N \| s_\theta)$ is the major source of computational complexity; it requires to solve the convex optimization problem (4.7) for all or selected values of the parameter θ given a particular value of the statistic $\bar{h}_N$.

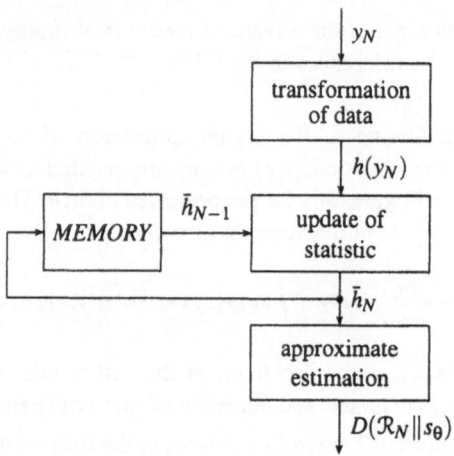

Fig. 5.1. Conceptual structure of estimation algorithm for independent observations.

Functional Approximation. The map

$$\boxed{(\bar{h}_N, \theta) \mapsto D(\mathcal{R}_N \| s_\theta)} \tag{5.5}$$

assigns to the vectors $\bar{h}_N \in \mathbb{R}^n$ and $\theta \in \mathcal{T} \subset \mathbb{R}^{\dim \theta}$ a scalar value of the K-L distance $D(\mathcal{R}_N \| s_\theta)$. Taken from the abstract point of view, the heart of the estimation algorithm is thus a scalar function of (highly) multivariate argument. In general, the function cannot be expressed analytically. We can, however, approximate it with another function that admits a parametric representation of sufficiently low dimension.

The problem to match a given function of multivariate argument with another, simpler one is well known in mathematics and engineering science, for instance,

- in *numerical mathematics* as multivariate functional approximation (de Boor and Rice, 1979; Light and Cheney, 1985),

- in *statistics* as nonparametric multiple regression (Friedman and Stuetzle, 1981; Huber, 1981; Breiman et al., 1984; Donoho and Johnstone, 1989; Friedman, 1991),

- in *computer science and engineering* as learning of artificial neural networks (Barron and Barron, 1988; Poggio and Girosi, 1990; Barron, 1993).

The list of references as well as different approaches is far from being complete; it only illustrates that the field has been attracting enormous interest.

As an example, we can think of a nonlinear estimator implemented via a neural network where the network is used as a "computational engine" rather than a model of data. The network can be "learnt" before estimation using a sequence of values of (5.5) at particular points $(\bar{h}_N, \theta)$.

Controlled Dynamic Systems

Structure of Algorithm. The use of a fixed statistic $h(y,z)$ implies even for dynamic systems a fixed structure of the estimation algorithm. The core of approximate Bayesian estimation is again in data compression and approximate restoration of inaccuracy

$$\left(y^{N+m}, u^{N+m}\right) \mapsto \bar{h}_N \mapsto D(\mathcal{R}_N \| s_\theta).$$

The conceptual structure of the estimation algorithm is outlined in Fig. 5.2. Owing to the model dynamics, the structure is slightly extended. Except separate memories, all blocks in the algorithm are again only functions of their input arguments, without any dynamic elements.

(a) The latest data (y_N, u_N) are first used to compose the model statistic (y_N, z_N).

(b) The transform $(y_N, z_N) \mapsto h(y_N, z_N)$ is usually easy but if the canonical statistic $h(y,z)$ is defined through an integral transform of log-density $\log s_\theta(y|z)$, the step may require numerical integration.

(c) The map $\left(\bar{h}_{N-1}, h(y_N, z_N)\right) \mapsto \bar{h}_N$ is a trivial update of the statistic value.

(d) The map $\bar{h}_N \mapsto D(\mathcal{R}_N \| s_\theta)$ requires to solve the convex optimization problem (4.14) for all or selected values of the parameter θ given a particular value of the statistic $\bar{h}_N$.

Functional Approximation. The heart of the algorithm is again the map

$$\boxed{(\bar{h}_N, \theta) \mapsto D(\mathcal{R}_N \| s_\theta)} \tag{5.6}$$

that assigns to the vectors $\bar{h}_N \in \mathbb{R}^n$ and $\theta \in \mathcal{T} \subset \mathbb{R}^{\dim \theta}$ a scalar value of the K-L distance $D(\mathcal{R}_N \| s_\theta)$. The only but practically important difference is that the dimensions n and $\dim \theta$ of the input vectors are typically much higher for controlled dynamic systems. In other words, the problem is the same, but its dimensionality increased.

The idea of multivariate functional approximation of the map (5.6) applies to the case similarly as for independent observations.

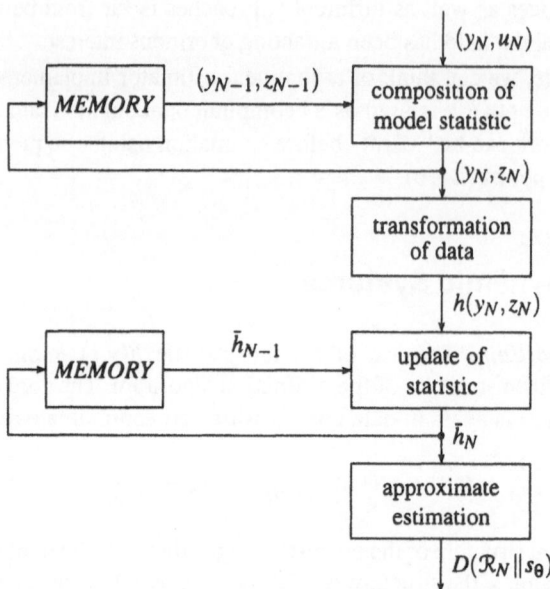

Fig. 5.2. Conceptual structure of estimation algorithm for controlled dynamic systems.

5.3 Parameter Tracking via Forgetting

In practice, the parameters θ of the underlying model are rarely constant, rather they vary slowly in time. The reason for model variations is either nonstationarity of the system behaviour itself, or the fact that the model considered captures the system behaviour only locally. Whatever the case, the estimation algorithm needs to be modified in order to become capable of tracking parameter changes.

The simplest and in practice most often used measure is that the older data are assigned a lesser weight in estimation compared with the newer data. We show how the forgetting of old information can be implemented in our general scheme.

Independent Observations

A sufficient description of data for any model of independent observations is given by the pair (N, r_N) where $N > 0$ is the number of observations and $r_N(y) \geq 0$ is the empirical density defined as

$$r_N(y) \triangleq \frac{1}{N} \sum_{k=1}^{N} \delta_N(y) \geq 0$$

with a simplified notation $\delta_N(y) = \delta(y - y_N)$.

We define two operations over the set of all such pairs

$$(N_1, r_{N_1}) + (N_2, r_{N_2}) = \left(N_1 + N_2, \frac{N_1 \, r_{N_1} + N_2 \, r_{N_2}}{N_1 + N_2} \right),$$

$$\lambda(N, r_N) = (\lambda N, r_N).$$

Using the above operations, the update of (N, r_N) with the new data $(1, \delta_{N+1})$ can be described in the following compact form

$$(N + 1, r_{N+1}) = (N, r_N) + (1, \delta_{N+1}). \tag{5.7}$$

Simple Forgetting. Introducing forgetting as a simple discounting of the number of observations, we get a modified algorithm

$$\boxed{(\nu_{N+1}, r_{N+1}) = \lambda(\nu_N, r_N) + (1, \delta_{N+1}),} \tag{5.8}$$

or, equivalently,

$$\nu_{N+1} = \lambda \nu_N + 1,$$

$$r_{N+1} = \frac{\lambda \nu_N}{\nu_{N+1}} r_N + \frac{1}{\nu_{N+1}} \delta_{N+1}.$$

The recursion starts from $\nu_0 = 0$. The forgetting factor is chosen as $0 < \lambda < 1$. Note that with $\lambda = 1$, the algorithm (5.8) coincides with the original one (5.7).

The batch form of (5.8) looks as follows

$$\nu_{N+1} = \sum_{k=1}^{N+1} \lambda^{N-k+1} = \frac{1 - \lambda^{N+1}}{1 - \lambda},$$

$$r_{N+1} = \frac{1}{\nu_{N+1}} \sum_{k=1}^{N+1} \lambda^{N-k+1} \delta_k.$$

Note that $\nu_N \to \frac{1}{1-\lambda}$ as $N \to \infty$. As a result of forgetting, the empirical density is a *time-discounted* mixture of Dirac functions pointing at particular data points.

Including Prior Information. When prior information about the observed data is available in the form (ν_0, r_0) with $\nu_0 > 0$ and a prior empirical density $r_0(y)$, we can incorporate it into (5.8) simply by changing the initial conditions. The modified recursion

$$\boxed{(\nu_{0,N+1}, r_{0,N+1}) = \lambda(\nu_{0,N}, r_{0,N}) + (1, \delta_{N+1})} \tag{5.9}$$

starts then from $\nu_{0,0} = \nu_0$ and $r_{0,0} = r_0$.

Equivalently, (5.9) can be written as

$$\nu_{0,N+1} = \lambda \nu_{0,N} + 1,$$

$$r_{0,N+1} = \frac{\lambda \nu_{0,N}}{\nu_{0,N+1}} r_{0,N} + \frac{1}{\nu_{0,N+1}} \delta_{N+1}.$$

The batch form of (5.9) looks as follows

$$v_{0,N+1} = \lambda^{N+1} v_0 + \sum_{k=1}^{N+1} \lambda^{N-k+1} = \lambda^{N+1} v_0 + \frac{1-\lambda^{N+1}}{1-\lambda},$$

$$r_{0,N+1} = \lambda^{N+1} \frac{v_0}{v_{0,N+1}} r_0 + \frac{1}{v_{0,N+1}} \sum_{k=1}^{N+1} \lambda^{N-k+1} \delta_k.$$

Since $\lambda^{N+1} \to 0$ as $N \to \infty$, $v_{0,N}$ and $r_{0,N}$ asymptotically coincide with v_N and r_N, respectively. In other words, the prior information is lost as a result of forgetting.

Preserving Prior Information. A simple modification of the formula (5.9) ensures that the prior information is preserved through forgetting. Just set

$$\boxed{(v_{0,N+1}, r_{0,N+1}) = \lambda (v_{0,N}, r_{0,N}) + (1-\lambda)(v_0, r_0) + (1, \delta_{N+1})} \qquad (5.10)$$

with $(v_{0,0}, r_{0,0})$ starting from (v_0, r_0). While in (5.9) we simply put less weight on all the previous data, in (5.10) we modify even the empirical density $r_{0,N}(y)$ so to make it closer to the prior empirical density $r_0(y)$.

The recursion (5.10) can be rewritten as follows

$$v_{0,N+1} = \lambda v_{0,N} + (1-\lambda) v_0 + 1,$$

$$r_{0,N+1} = \frac{\lambda v_{0,N}}{v_{0,N+1}} r_{0,N} + \frac{(1-\lambda) v_0}{v_{0,N+1}} r_0 + \frac{1}{v_{0,N+1}} \delta_{N+1}.$$

From here, the batch form of (5.10) easily follows

$$v_{0,N+1} = v_0 + \sum_{k=1}^{N+1} \lambda^{N-k+1}$$

$$= v_0 + v_{N+1}$$

$$= v_0 + \frac{1-\lambda^{N+1}}{1-\lambda},$$

$$r_{0,N+1} = \frac{v_0}{v_{0,N+1}} r_0 + \frac{1}{v_{0,N+1}} \sum_{k=1}^{N+1} \lambda^{N-k+1} \delta_k$$

$$= \frac{v_0}{v_0 + v_{N+1}} r_0 + \frac{v_{N+1}}{v_0 + v_{N+1}} \frac{1}{v_{N+1}} \sum_{k=1}^{N+1} \lambda^{N-k+1} \delta_k$$

Note that $v_{0,N} \to v_0 + \frac{1}{1-\lambda}$ as $N \to \infty$. The incorporation of prior information is reflected by the fact that the empirical density $r_{0,N+1}$ is now a *weighted mixture* of the prior density r_0 and a time-discounted mixture of Dirac functions pointing at particular data points. The corresponding weights are $\frac{v_0}{v_0 + v_{N+1}}$ and $\frac{v_{N+1}}{v_0 + v_{N+1}}$, respectively.

Effect of Forgetting on Posterior Density. Provided the prior density of the unknown parameter is chosen in the conjugate form

$$p_0(\theta) \propto \exp(-v_0 K(r_0 : s_\theta)),$$

the posterior density is of the same form

$$p_N(\theta) \propto \exp\left(-v_{0,N} K(r_{0,N}:s_\theta)\right).$$

It is easy to see that the forgetting according to

$$(v_{0,N}, r_{0,N}) \mapsto \lambda(v_{0,N}, r_{0,N})$$

results simply in a more flattened posterior density

$$\tilde{p}_N(\theta) \propto \exp\left(-\lambda v_{0,N} K(r_{0,N}:s_\theta)\right)$$
$$\propto \left(p_N(\theta)\right)^\lambda$$

and coincides thus with the classical exponential forgetting.
In contrast, the use of modified forgetting

$$(v_{0,N}, r_{0,N}) \mapsto \lambda(v_{0,N}, r_{0,N}) + (1 - \lambda)(v_0, r_0)$$

preserves the prior density through the recursion

$$\tilde{p}_N(\theta) \propto \exp\left(-\lambda v_{0,N} K(r_{0,N}:s_\theta)\right) \exp\left(-(1-\lambda) v_0 K(r_0:s_\theta)\right)$$
$$\propto \left(p_N(\theta)\right)^\lambda \left(p_0(\theta)\right)^{1-\lambda}.$$

Evaluation of Statistic. Owing to the form of the empirical density update, it is very easy to evaluate the value of the statistic

$$\bar{h}_N = \int r_N(y) h(y) \, dy.$$

With the simple forgetting (5.8), the update of $(v_N, \bar{h}_N)$ is given by

$$v_{N+1} = \lambda v_N + 1,$$
$$\bar{h}_{N+1} = \frac{\lambda v_N}{v_{N+1}} \bar{h}_N + \frac{1}{v_{N+1}} h(y_{N+1}),$$

starting from $v_0 = 0$.
With prior information included, the statistic value $\bar{h}_{0,N+1}$ is affected by the prior value

$$\bar{h}_0 = \int r_0(y) h(y) \, dy.$$

The forgetting scheme (5.9) implies the following update of $(v_{0,N}, \bar{h}_{0,N})$

$$v_{0,N+1} = \lambda v_{0,N} + 1,$$
$$\bar{h}_{0,N+1} = \frac{\lambda v_{0,N}}{v_{0,N+1}} \bar{h}_{0,N} + \frac{1}{v_{0,N+1}} h(y_{N+1})$$

while using (5.10) we have

$$v_{0,N+1} = \lambda v_{0,N} + (1 - \lambda) v_0 + 1,$$
$$\bar{h}_{0,N+1} = \frac{\lambda v_{0,N}}{v_{0,N+1}} \bar{h}_{0,N} + \frac{(1 - \lambda) v_0}{v_{0,N+1}} \bar{h}_0 + \frac{1}{v_{0,N+1}} h(y_{N+1}).$$

Both recursions start from $v_{0,0} = v_0$ and $\bar{h}_{0,0} = \bar{h}_0$.

Controlled Dynamic Systems

All what was said above for the case of independent observations extends straight-forwardly to the case of controlled dynamic systems. The only change is that the empirical density $r_N(y)$ is substituted by

$$r_N(y,z) = \frac{1}{N} \sum_{k=m+1}^{N+m} \delta_k(y) \geq 0$$

with a simplified notation $\delta_k(y,z) = \delta(y - y_k, z - z_k)$. In addition, all the sums $\sum_{k=1}^{N+1}$ are replaced by $\sum_{k=m+1}^{N+m+1}$, in other words, the time indices are lifted by the smallest integer $m > 0$ for which z_{m+1} is defined.

5.4 Historical Notes

Iterative Solution of Minimum K-L Distance Problem. The convergence of the iterative solution to the optimum was proved for discrete distributions by Ireland and Kullback (1968) and for continuous distributions by Kullback (1968). A general and rigorous treatment of convergence, existence and uniqueness can be found in Csiszár (1975), along with discussion on the connection between the iterative minimum K-L distance projection and the iterative proportional fitting procedure. Burr (1989) described an application of the iterative projection algorithm in the minimum K-L distance spectral analysis (Shore, 1981). Dykstra (1985) proposed an iterative procedure that only requires that the sets of distributions determined by particular constraints are convex.

Exponential Forgetting. Exponential forgetting—weighting data according to their "age" is a classical technique of making estimation adaptive; a good overview can be found in Jazwinski (1966). Peterka (1981) showed the Bayesian interpretation of exponential forgetting as flattening of the posterior density. A regularized version of exponential forgetting was derived and analysed using K-L distance in Kulhavý and Zarrop (1993) and Kulhavý and Kraus (1996).

6. Concluding Remarks

The strong point of approximation developed in Chap. 3 and 4 is that it provides *at the same time*

(a) guidelines for design of proper data compression given a particular model class,

(b) method of approximate restoration of the likelihood function and the posterior density from compressed data.

There are, of course, numerous connections between the scheme proposed here and results known in statistics and system identification. Many of the connections were already pointed out in the historical notes in the preceding chapters.

The aim of the following remarks is to stress those features of the proposed solution that distinguish it from others. In the end, we indicate major open problems that wait for more investigation.

6.1 Summary of Key Points

Bayesian Versus Point Estimation. Design of approximate estimation is much harder in the Bayesian case than it is in point estimation. While in point estimation we are interested only in finding a certain *point* in the parameter space, in Bayesian estimation we compute a whole *function* of the unknown parameter—likelihood, posterior density or perhaps inaccuracy as in our case. Bayesian estimation is thus in general much more costly. The higher cost is justified when an accurate description of the parameter uncertainty is required like in robust control design or basically in any decision-making based on moderate-size samples.

The view of Bayesian estimation in the preceding chapters was centred fully around the computational issues. Very little space was devoted to discussion of the pros and cons of the Bayesian paradigm. Very little was said about the role of prior information. The problem solved here was rather a practical one—to propose approximation of inaccuracy/likelihood/posterior that would be well justified and broadly applicable.

Estimation Viewed as Approximation. Mathematical statistics views estimation usually as a kind of (ill-posed) inverse problem. The classical statistical approach is based on the following premises:

(a) the data are assumed to appear by sampling from a probability distribution that belongs to a known family;

(b) everything is 'modelled', meaning that a joint distribution (typically parametrized by some unknown parameters) of all the observed data can be composed;

(c) except degenerate cases, the true estimate can be found only asymptotically.

In contrast to the abstract mathematical theory, the solution of practical problems calls for substantial relaxation of the assumptions:

(a) the true mechanism generating data is usually too complex to be completely described by any model of admissible complexity, hence the model considered is only approximation of the true system behaviour;

(b) only the system response, i.e., the dependence of the current output on the past outputs, inputs and possibly external disturbances is typically modelled while there is no explicit model for the inputs and (measured) disturbances to the system;

(c) the amount of data is, as a rule, limited.

To keep up with the latter situation, it is better to transpose estimation into the form of an explicit approximation problem. Not only it gives us more insight into what we do, it also facilitates further approximation which may be necessary to make the solution feasible under given constraints on computational memory and time. The view of estimation as approximation leads us to prefer using information measures—inaccuracy and K-L distance to the notions of likelihood and posterior density.

From Data Matching to Probability Matching. The non-probabilistic methods of estimation typically match the sequence of observed data with another, model-based sequence so to minimize some measure of distance between both. An analogous picture applies to probability-based estimation as well; the objective is to match the empirical distribution of data with a theoretical, model-based distribution so to minimize a kind of "distance" between both again.

The formal similarity between matching data and matching probabilities is pointed out in Table 6.1.

Use of Inaccuracy and Kullback-Leibler Distance. In point estimation, much less is required from the estimator so that we have considerably more freedom in its construction. Thus, if we are interested in point estimates and its asymptotic properties, we are naturally led to try various "distances" in the space of probability distributions. The class of f-divergences introduced by Csiszár (1967) and Ali and Silvey (1966) provides a general enough framework for this.

Since the objective of Bayesian estimation is to reconstruct the whole function of the unknown parameter, the use of inaccuracy and K-L distance seems to be the only choice if we are to end up with results close to what the rules of probability theory do ideally for us.

Table 6.1. Formal correspondence of some notions of the Euclidean geometry of data and Pythagorean geometry of probability distributions of data.

Data Matching	Probability Matching
vector of data	distribution of data
hyperplane	exponential family
Euclidean distance between the true and model sequences	inaccuracy of empirical distribution relative to theoretical distribution
Euclidean distance between points in a hyperplane	Kullback-Leibler distance between distributions in exponential family

Pythagorean Geometry. The key tool used in our approach is the Pythagorean relationship that enables us to decompose the inaccuracy of the empirical density relative to a theoretical density into a sum of two terms one of which can be made, by a proper choice of statistic, almost independent of the unknown parameter. The Pythagorean relationship is well known in this context in the statistical literature, namely for the case of independent and discrete observations. Here the idea of similar decomposition was extended to controlled dynamic systems and continuous data.

Let us stress that the Pythagorean approximation assumes neither stationarity of the underlying processes, nor complete description of inputs and disturbances to the system. The "trick" is that the unknown distribution of data entering the condition of the theoretical density is *estimated* explicitly using available information rather than reconstructed asymptotically.

It is a pleasant consequence of using a unified methodology that the solutions for independent observations and controlled dynamic systems have identical structure.

Information Inequalities. The major benefit of using the information measures is the existence of meaningful bounds on the minimum K-L distance both for independent observations

$$0 \leq D(\mathcal{R}_N \| s_\theta) \leq D(r \| s_\theta), \quad r \in \mathcal{R}_N$$

and controlled dynamic systems

$$-\max_{r \in \mathcal{R}_N} H(\tilde{r}) \leq D(\mathcal{R}_N \| s_\theta) \leq D(r \| s_\theta), \quad r \in \mathcal{R}_N.$$

Together with the (anti)monotonicity property

$$D(\mathcal{R}_N \| s_\theta) \leq D(\mathcal{R}'_N \| s_\theta) \quad \text{for} \quad \mathcal{R}_N \supset \mathcal{R}'_N,$$

these inequalities let us know what are the best and worst possible outcomes of approximation and what we lose or gain when the set $\mathcal{R}_N$ varies as a result of the statistic change.

Beating Computational Inaccuracy. The optimum solution to estimation with compressed data would be to compute the probability of observing a given value of the statistic, for every value of the unknown parameter. The probability is found by integration of the joint density over a subset of its domain which is a space of high dimension, growing with the number of observations. The task is clearly infeasible for general parametric families.

The approximation suggested replaces the explicit integration by solving a convex optimization problem. This problem also includes integration but only over the space of a finite dimension, namely the dimension of data entering the model, i.e., Y or (Y,Z).

Recursive Character of Approximation. The statistic considered throughout the book is of the form

$$\bar{h}_N = \frac{1}{N} \sum_{k=1}^{N} h(y_k)$$

for independent observations and

$$\bar{h}_N = \frac{1}{N} \sum_{k=m+1}^{N+m} h(y_k, z_k)$$

for controlled dynamic systems where $h(y)$ and $h(y,z)$ are fixed vector statistics of single observation Y and (Y,Z), respectively. Obviously, the sample average can easily be computed recursively. The only information about data we use to restore the likelihood function is the statistic value $\bar{h}_N$ and the number of observations N. As a result, the approximation scheme is well suited for recursive implementation.

Global Versus Local Methods. The recursive algorithms used for point estimation usually do not use a fixed statistic. Instead, the statistic is adjusted so to be nearly optimal for the last estimate. When the estimate gets close to the true or optimum parameter value, the algorithms are able, through a fine adjustment of the statistic, to ensure asymptotic convergence of the estimate towards the ideal value.

The local methods thus ensure (under some regularity assumptions) consistency of point estimates. As a rule, they provide also information about the shape of likelihood in a small neighbourhood of the optimum estimate (typically utilizing the asymptotic normality of likelihood). They say very little, however, about the likelihood shape over larger regions of the parameter space.

In contrast, the approximation proposed in Chap. 4 can be viewed as a global method that is able, with a proper statistic, to locate the whole likelihood function within a set of possible likelihoods. The price we pay for the global character of approximation is that we cannot in general avoid a certain bias of point estimates defined through the approximate likelihood. The consistency is ensured only when the model family includes the true distribution of data and the statistic is rich enough to describe the distribution of data entering the condition of the theoretical density.

To sum up, if one is interested in point estimates only and is sure enough that his prior estimate (serving as the starting value for the algorithm) is close to the true or optimum point, he should use one of the local methods. If one wishes to estimate the whole likelihood, he will probably need a global method. The global method can also help to accelerate the initial convergence of point estimation.

Needless to say, there is much room here for more or less ingenious combinations of global and local methods.

The Value of Prior Information. Any estimation that uses compressed data not sufficient for a given model is necessarily a compromise between the dimension of the data statistic and the precision of approximation. Estimation for a complex model may require statistics of very high dimensions if a good agreement between the true and approximate likelihoods is to be achieved for all possible parameter values.

The only escape from this dilemma is to use as much prior information as possible. Having enough prior information about the unknown parameters, i.e., about the actual distribution of observed data, we can construct the statistic so to ensure a good agreement between the true and approximate likelihoods only over the relevant subset of the parameter space. Such a statistic may have a considerably smaller dimension.

Symmetric Role of Data and Model. The data and model, represented by the empirical and theoretical densities, play in the approximation essentially symmetric roles. Hence, we can rather freely change the model class *during estimation* provided the statistic is chosen so to bring enough information for estimation of all models tried. This feature opens quite interesting possibilities for system identification.

(a) *Adaptive identification.* Instead of estimating parameters of a big, complex model that covers all potentially possible system responses, we can build the model class step by step, extending the current model or looking for a better model in the "neighbourhood" of the current model.

(b) *Robust identification.* To account for possible mismodelling, we can consider models $s_{\theta,\xi}$ with nuisance parameters $\xi \in \mathcal{X}$. Instead of estimating explicitly the nuisance parameters, we can compute the minimum K-L distance $D(\mathcal{R}_N \| \mathcal{S}_\theta)$ where $\mathcal{S}_\theta = \{s_{\theta,\xi} : \xi \in \mathcal{X}\}$.

(c) *"Bayesification" of point estimation.* Given a certain point estimate of the unknown parameter, we may pick an arbitrary set of parameter values in its neighbourhood and to calculate, given the statistic value only, the approximate likelihood over these points.

(d) *Analysis of mismodelling.* Considering various model classes $\mathcal{S}_i$, we can calculate the minimum K-L distance $D(\mathcal{R}_N \| \mathcal{S}_i)$ and use it as a measure of goodness of fit for particular model classes.

6.2 Open Problems

There are many points concerning the approximation that need to be clarified and analysed in more detail. Technical questions include analysis of conditions under which solutions to the optimization problems exist, and possible extensions to problems with weaker regularity assumptions. Conceptually, there are three fundamental issues here.

Approximation of Minimum K-L Distance. The Pythagorean approximation requires to solve the same optimization problem for every possible value of the unknown parameter. Although usually it suffices to solve the problem for a limited set of points of interest and then possibly interpolate between them, one naturally tends to prefer a solution that returns the approximate inaccuracy/likelihood/posterior as an explicit function of the parameter.

The last is impossible in general; given a complex enough model, the minimum K-L distance as a function of the unknown parameter is too complex to be expressed using any finitely-parametrized function. This fact calls for further approximation of the solution proposed.

Given the meaning of K-L distance as a measure of information for discrimination between probability distributions, it is natural to require that any approximation of K-L distance is bounded from above by the true K-L distance. It formalizes the obvious fact that the amount of information for discrimination never increases as a result of approximation. Using a purely numerical approximation, this monotonicity requirement may easily be violated. Hence, schemes of approximation that preserve the monotonicity are wanted.

Optimization of Statistic. We have pointed out repetitively the importance of a proper choice of the data statistic. The class of necessary statistic introduced in Chap. 4 marks an area outside of which it makes little sense to move. Yet, even within the class of necessary statistics we are left with huge freedom concerning the choice of a particular statistic.

The natural goal is to make the necessary statistic almost sufficient, at least locally for those models s_θ, $\theta \in A \subset \mathcal{T}$ that are considered more likely. In terms of the Pythagorean geometry, we want to achieve the state when the minimum K-L distance projections of the model densities s_θ, $\theta \in A$ onto $\mathcal{R}_N$ almost coincide. If we succeed to construct an exponential family that locally, for $\theta \in A$, approximates the model family, then the canonical statistic of the approximating exponential family defines the single-data statistic to be used.

The geometric view makes it easy to quantify the deviation from the perfect sufficiency, e.g., by measuring the variability of projections $s_{\theta,\hat{\lambda}}$, $\theta \in A$ onto $\mathcal{R}_N$. In principle, it seems possible to formulate the choice of statistic given prior information as an explicit optimization problem.

Extension to Filtering. Bayesian estimation of a time-variable parameter θ_k whose evolution in time is described by the conditional density $p(\theta_{k+1}|\theta_k, y^k, u^k)$ represents a problem that has a more difficult structure than estimation of the constant parameter. In particular, the recursive calculation of the conditional density function $p(\theta_{k+1}|y^k, u^k)$ includes the convolution

$$p(\theta_{k+1}|y^k, u^k) = \int p(\theta_{k+1}|\theta_k, y^k, u^k)\, p(\theta_k|y^k, u^k)\, d\theta_k.$$

This breaks the simple product-like structure of the likelihood we utilized in Chap. 2 to introduce information measures.

Note, however, that the product-like structure is preserved through the joint density of all observations and *all parameters*, i.e., including the past ones. This fact suggests one possible approach to approximate filtering—treat the time-variable parameters as missing or unmeasurable data and estimate only the constant parameters of the underlying model and possibly some of the "missing data".

At any rate, the question whether even the Bayesian estimation of time-variable parameters can be transposed into the form of an explicit approximation problem using information measures remains to be a challenging problem.

6.3 Outlook

The use of abstract mathematical models in place of real systems and processes has a long history. It has become common in fields as different as physics, statistics, engineering, econometrics, or biology. In general, to build a faithful model of a given system requires a deep understanding of physical, chemical, biological, etc. processes underlying the system operation. A complete analytic solution is thus rather "expensive", measured by the time and qualification required.

It is one of the reasons why a lot of effort has been spent to automate the process of building a model based on *data* observed on the system. Fitting a parametric model to data is an attractive solution, especially as the relative cost of measurements and computations is steadily decreasing. This does not mean, of course, that having enough data and enough computational power, we do not need to care about the system. There are several features that make system identification a difficult and nontrivial task.

(a) *Data usually behave in an unpredictable manner.* This is a trivial observation that calls usually for the use of statistical methods of inference. Quite often, however, the source of 'randomness' is in the mismatch between a simple model and the complex reality rather than in the inherently stochastic behaviour of the plants and processes modelled.

(b) *Information contained in data is often considerably reduced before used for estimation.* In practice, data available for identification are rarely complete. In recursive estimation, data are compressed so that only the value of a certain data statistic is available. In economic and social applications, data are usually aggregated so that only sums or averages over a certain time period are recorded. Frequently some data items are completely missing. Measurements are often corrupted by noise or systematic errors.

(c) *The model class typically does not include the actual system.* As usual in science, to make predictions about the performance of system identification, a number of simplifying assumptions have to be postulated. The assumption that the true system belongs to a certain model class is commonly accepted. Strictly speaking, this is never true in reality.

(d) *The computer resources are always limited.* Only a finite and limited amount of computer memory and time is available for implementation of any identification algorithm. When the algorithm is too complex to be implemented in its theoretically optimal form, it has to be approximated in some way. The use of approximation increases in general the uncertainty of the results of inference.

None of known paradigms of inference seems to address all the points. In particular, probability-based inference often fails because of the extreme dimensionality of related computations. The use of information measures can then appear as a good alternative.

A. Selected Topics from Probability Theory

The purpose of this section is to remind the reader of basic operations of probability calculus and to make our presentation largely self-contained. The intention was to minimize the use of measure theory. The following summary follows essentially Grimmett and Stirzaker (1992); other good references on this level are Papoulis (1991) or Shiryaev (1984).

A.1 Probability Spaces

Events. Consider an experiment or trial. The set of all possible outcomes of an experiment is called the sample space and is denoted by Ω. More complex combinations of outcomes are called events. Events can be seen as subsets of Ω.

Not all the subsets of Ω need to be events. A collection of subsets of Ω is only required to be closed under the operations of taking countable unions and countable intersections. Any collection of subsets of Ω with these properties is called σ-algebra. More formally, a collection of subsets of Ω is called a σ-algebra if it satisfies the following conditions:

(a) $\varnothing \in \mathscr{F}$,

(b) if $A_1, A_2, \ldots \in \mathscr{F}$ then $\bigcup_{i=1}^{\infty} A_i \in \mathscr{F}$,

(c) if $A \in \mathscr{F}$ then $\bar{A} \in \mathscr{F}$

where $\bar{A}$ denotes the complement of a subset A of Ω.

Probability Measure. A probability measure P on $(\Omega, \mathscr{F})$ is a function $P : \mathscr{F} \to [0,1]$ satisfying

(a) $P(\varnothing) = 0, \quad P(\Omega) = 1$;

(b) if $A_1, A_2, \ldots$ is a collection of disjoint members of $\mathscr{F}$, so that $A_i \cap A_j = \varnothing$ for all pairs i, j satisfying $i \neq j$, then

$$P\left(\bigcup_{i=1}^{\infty} A_i \right) = \sum_{i=1}^{\infty} P(A_i).$$

The property (b) says that P is a countably additive function.

Probability Space. The triple $(\Omega,\mathscr{F},P)$, comprising of a set Ω, a σ-algebra $\mathscr{F}$ of subsets of Ω, and a probability measure P on $(\Omega,\mathscr{F})$, is called a probability space.

Product Spaces. Suppose two experiments with associated probability spaces $(\Omega_1,\mathscr{F}_1,P_1)$ and $(\Omega_2,\mathscr{F}_2,P_2)$, respectively. The sample space of the pair of experiments, considered jointly, is the collection

$$\Omega_1 \times \Omega_2 = \{(\omega_1,\omega_2) : \omega_1 \in \Omega_1, \omega_2 \in \Omega_2\}$$

of ordered pairs. The appropriate σ-algebra of events is constructed as the smallest σ-algebra $\mathscr{G}$ of subsets of $\Omega_1 \times \Omega_2$ which contains $\mathscr{F}_1 \times \mathscr{F}_2$.

There are many measures that can be applied to $(\Omega_1 \times \Omega_2,\mathscr{G})$. The simplest one is the product measure defined through

$$P_{12}(A_1 \times A_2) = P_1(A_1)P_2(A_2)$$

for any $A_1 \in \mathscr{F}_1$ and $A_2 \in \mathscr{F}_2$. It can be shown that the measure can be extended from $\mathscr{F}_1 \times \mathscr{F}_2$ to the whole of $\mathscr{G}$. The resulting probability space $(\Omega_1 \times \Omega_2,\mathscr{G},P_{12})$ is called the *product space* of $(\Omega_1,\mathscr{F}_1,P_1)$ and $(\Omega_2,\mathscr{F}_2,P_2)$.

Products of larger numbers of spaces are constructed similarly.

Conditional Probability. Consider two events A and B. If $P(B) > 0$ then the conditional probability that A occurs given that B is defined to be

$$P(A|B) = \frac{P(A \cap B)}{P(B)}. \tag{A.1}$$

Total Probability. Let $A_1,A_2,\ldots,A_n$ be a partition of Ω, i.e.,

$$A_i \cap A_j = \varnothing \quad \text{when} \quad i \neq j, \quad \text{and} \quad \bigcup_{i=1}^{n} A_i = \Omega.$$

Let $P(A_i) > 0$ for each $i = 1,2,\ldots,n$. Then

$$P(B) = \sum_{i=1}^{n} P(B|A_i)P(A_i). \tag{A.2}$$

In particular, if $0 < P(A) < 1$, then

$$P(B) = P(B|A)P(A) + P(B|\bar{A})P(\bar{A}). \tag{A.3}$$

Bayes's Theorem. For any events A and B with $P(A) > 0$

$$P(A \cap B) = P(B|A)P(A). \tag{A.4}$$

This formula, called *chain rule* or multiplication formula for probabilities, easily extends to families of events $A_1,A_2,\ldots,A_n$ such that $P(A_1 \cap \ldots \cap A_{n-1}) > 0$

$$P(A_1 \cap \ldots \cap A_n) = P(A_1) P(A_2|A_1) \ldots P(A_n|A_1 \cap \ldots \cap A_{n-1}). \tag{A.5}$$

Similarly, for any events A and B with $P(B) > 0$

$$P(A \cap B) = P(A|B) P(B). \tag{A.6}$$

If both $P(A) > 0$ and $P(B) > 0$, (A.4) and (A.6) yield *Bayes's formula*

$$P(A|B) = \frac{P(B|A) P(A)}{P(B)}. \tag{A.7}$$

Let $A_1, A_2, \ldots, A_n$ be a partition of Ω with $P(A_i) > 0$ for each $i = 1, 2, \ldots, n$. Then the formulae (A.2) and (A.7) imply together *Bayes's theorem*

$$P(A_i|B) = \frac{P(B|A_i) P(A_i)}{\sum_{j=1}^{n} P(B|A_j) P(A_j)}. \tag{A.8}$$

Independence. Two events A and B are called *independent* if

$$P(A \cap B) = P(A) P(B).$$

More generally, a family $\{A_i : i \in I\}$ is called *independent* if

$$P\left(\bigcap_{i \in J} A_i\right) = \prod_{i \in J} P(A_i)$$

for all finite subsets J of I.

Let C be an event with $P(C) > 0$. Two events A and B are called *conditionally independent* given C if

$$P(A \cap B|C) = P(A|C) P(B|C).$$

Conditional independence naturally extends to families of events.

Random Variables. One is often interested in consequences of the outcome of an experiment rather than in the experiment itself. Such consequences, when real valued, may be thought of as functions which map Ω into the real line $\mathbb{R}$. These functions are called 'random variables'.

More precisely, a *random variable* is a function $X: \Omega \to \mathbb{R}$ with the property that $\{\omega \in \Omega : X(\omega) \leq x\} \in \mathscr{F}$ for each $x \in \mathbb{R}$. Any X with this property is called $\mathscr{F}$-measurable.

The *distribution function* of a random variable X is the function $F: \mathbb{R} \to [0, 1]$ given by

$$F(x) = P(X \leq x).$$

The assumption that X is $\mathscr{F}$-measurable ensures that the events

$$\{X \leq x\} = \{\omega \in \Omega : X(\omega) \leq x\},$$

for each $x \in \mathbb{R}$, are assigned values by a probability measure P.

It is a common convention to use upper case letters, like X, Y, Z, for random variables, and to reserve lower-case letters, like x, y, z, for numerical values of these variables.

A.2 Discrete Random Variables

The random variable is called *discrete* if it takes values in a countable subset $\{y_1, y_2, \ldots\}$ of $\mathbb{R}$. It is easy to see that the distribution function $F(x)$ of a discrete random variable has jump discontinuities at the values $y_1, y_2, \ldots$ and is constant in between.

Mass Function. The *(probability) mass function* of a discrete random variable X is the function $f : \mathbb{R} \to [0, 1]$ given by $f(x) = P(X = x)$. The distribution and mass functions are related to each other by

$$F(x) = \sum_{i : y_i \le x} f(y_i), \quad f(x) = F(x) - \lim_{y \uparrow x} F(y).$$

The probability mass function $f : \mathbb{R} \to [0, 1]$ satisfies

(a) $f(x) \ne 0$ if and only if x belongs to some countable set $\{y_1, y_2, \ldots\}$,

(b) $\sum_i f(y_i) = 1$.

Joint Mass Function. The *joint distribution function* $F : \mathbb{R}^2 \to [0, 1]$ of X and Y, where X and Y are discrete variables, is given by

$$F(x, y) = P(X \le x, Y \le y).$$

Their *joint (probability) mass function* $f : \mathbb{R}^2 \to [0, 1]$ is defined as

$$f(x, y) = P(X = x, Y = y).$$

We write $F_{X,Y}$ and $f_{X,Y}$ when we need to stress the role of X and Y. Joint distribution functions and joint probability mass functions of larger collections of variables are defined similarly.

Marginal Mass Function. Suppose that X and Y have a joint probability mass function $f_{X,Y}$. It follows from

$$f_X(x) = P(X = x) = \sum_y P(X = x, Y = y)$$

that the *marginal mass function* f_X of X is

$$f_X(x) = \sum_y f_{X,Y}(x, y). \tag{A.9}$$

Similarly, the *marginal mass function* f_Y of Y is

$$f_Y(y) = \sum_x f_{X,Y}(x, y). \tag{A.10}$$

Conditional Mass Function. The conditional distribution function of Y given $X = x$, written $F_{Y|X}(\cdot|x)$, is defined by

$$F_{Y|X}(y|x) = P(Y \leq y|X = x)$$

for any x such that $P(X = x) > 0$. The conditional probability mass function of Y given $X = x$, written $f_{Y|X}(\cdot|x)$, is defined by

$$f_{Y|X}(y|x) = P(Y = y|X = x)$$

for any x such that $P(X = x) > 0$. Taking into account (A.1), we have

$$f_{Y|X}(y|x) = \frac{f_{X,Y}(x,y)}{f_X(x)} \tag{A.11}$$

for any x such that $f_X(x) > 0$. Conditional distribution and mass functions are undefined at values of x for which $P(X = x) = 0$.

Bayes's Theorem. The chain rule for mass functions follows directly from the definition (A.11)

$$f_{X,Y}(x,y) = f_{Y|X}(y|x)\, f_X(x). \tag{A.12}$$

Quite analogously, we have

$$f_{X,Y}(x,y) = f_{X|Y}(x|y)\, f_Y(y). \tag{A.13}$$

Taking (A.12) and (A.13) together, we obtain *Bayes's theorem* for mass functions

$$f_{X|Y}(x|y) = \frac{f_{Y|X}(y|x)\, f_X(x)}{\sum_x f_{Y|X}(y|x)\, f_X(x)}. \tag{A.14}$$

Independence. X and Y are *independent*, denoted as $X \perp Y$, if the events $\{X = x\}$ and $\{Y = y\}$ are independent for all x and y. In terms of mass functions, it reads

$$f_{X,Y}(x,y) = f_X(x)\, f_Y(y) \quad \text{for all } x,y \in \mathbb{R}. \tag{A.15}$$

Note that X and Y are independent if and only if $f_{Y|X} = f_Y$.

X and Y are *conditionally independent*, denoted as $X \perp Y|Z$, given another variable Z if the events $\{X = x\}$ and $\{Y = y\}$ are conditionally independent for all x and y given $\{Z = z\}$. This definition is equivalent to

$$f_{X,Y|Z}(x,y|z) = f_{X|Z}(x|z)\, f_{Y|Z}(y|z) \quad \text{for all } x,y \in \mathbb{R}. \tag{A.16}$$

The conditional mass functions are defined for any z such that $f_Z(z) > 0$.

Expectation. The *mean value*, or *expectation*, or *expected value* of a discrete variable X with mass function f is defined to be

$$E(X) = \sum_{x:f(x)>0} x f(x)$$

whenever this sum is absolutely convergent. For notational convenience, we may write $E(g(X)) = \sum_x g(x) f(x)$ as all but countably many of contributions in the sum are zero.

If X has a mass function f and $g: \mathbb{R} \to \mathbb{R}$, then

$$E(g(X)) = \sum_x g(x) f(x) \tag{A.17}$$

whenever this sum is absolutely convergent. Note that this way of computing $E(g(X))$ is much easier than to calculate directly $E(Y)$ of $Y = g(X)$.

The *conditional expectation* of Y given $X = x$ is defined by

$$E(Y|X = x) = \sum_y y f_{Y|X}(y|x). \tag{A.18}$$

Since the conditional expectation depends on x taken by X, we can think of it as function of X itself—the conditional expectation of Y given X, written as $E(Y|X)$.

The conditional expectation $E(Y|X)$ satisfies

$$E(E(Y|X)) = E(Y).$$

A.3 Continuous Random Variables

The random variable is called *continuous* if its distribution function can be expressed as

$$F(x) = \int_{-\infty}^{x} f(u)\, du \quad x \in \mathbb{R} \tag{A.19}$$

for some integrable function $f: \mathbb{R} \to [0, \infty)$. The distribution function of a continuous random variable is necessarily an absolutely continuous function.

Density Function. The function f in (A.19) is called the *(probability) density function* of X. The density function is not uniquely determined by (A.19) since two integrable functions which take identical values except at some specific point have the same integrals. However, if F is differentiable at x then we normally set

$$f(x) = \frac{d}{dx} F(x).$$

If X has a density function f, then

(a) $\int_{-\infty}^{\infty} f(x)\, dx = 1$,

(b) $P(X = x) = 0$ for all $x \in \mathbb{R}$,

(c) $P(a \le X \le b) = \int_a^b f(x)\,dx.$

The formula (c) has a more general counterpart. Let $\mathscr{I}$ be the collection of open intervals in $\mathbb{R}$. $\mathscr{I}$ can be extended to a unique smallest σ-algebra $\mathscr{B}$ which contains $\mathscr{I}$; one can take intersection of all σ-algebras that contain $\mathscr{I}$. $\mathscr{B}$ is called *Borel σ-algebra* and contains *Borel sets*. For all Borel sets $B \in \mathscr{B}$

$$P(X \in B) = \int_B f(x)\,dx. \tag{A.20}$$

A function $f : \mathbb{R} \to \mathbb{R}$ is said to be *Borel measurable* if $f^{-1}(B) \in \mathscr{B}$ for all $B \in \mathscr{B}$.

Joint Density Function. The *joint distribution function* of continuous variables X and Y is the function $F : \mathbb{R}^2 \to [0,1]$ given by

$$F(x,y) = P(X \le x, Y \le y).$$

X and Y are *(jointly) continuous* with *joint (probability) density function* $f : \mathbb{R}^2 \to [0,1]$ if

$$F(x,y) = \int_{-\infty}^{y} \int_{-\infty}^{x} f(u,v)\,du\,dv \quad \text{for each } x,y \in \mathbb{R}.$$

We write $F_{X,Y}$ and $f_{X,Y}$ when we need to stress the role of X and Y. Joint distribution functions and joint probability density functions of larger collections of variables are defined similarly.

Marginal Density Function. Consider continuous variables X and Y with joint distribution function $F_{X,Y}$ and joint density function $f_{X,Y}$. The *marginal distribution functions* of X and Y are then

$$F_X(x) = P(X \le x) = F_{X,Y}(x,\infty), \quad F_Y(y) = P(Y \le y) = F_{X,Y}(\infty,y)$$

where $F(x,\infty) = \lim_{y\to\infty} F(x,y)$ and $F(\infty,y) = \lim_{x\to\infty} F(x,y)$. From

$$F_X(x) = \int_{-\infty}^{x} \left(\int_{-\infty}^{\infty} f(u,y)\,dy \right) du$$

it follows that the *marginal density function* of X is

$$f_X(x) = \int_{-\infty}^{\infty} f(x,y)\,dy. \tag{A.21}$$

Similarly, the *marginal density function* of Y is

$$f_Y(y) = \int_{-\infty}^{\infty} f(x,y)\,dx. \tag{A.22}$$

Conditional Density Function. The *conditional distribution function* of Y given $X = x$, written $F_{Y|X}(\cdot|x)$, is defined to be

$$F_{Y|X}(y|x) = P(Y \leq y|X = x) = \int_{-\infty}^{y} \frac{f_{X,Y}(x,v)}{f_X(x)}\, dv$$

for any x such that $f_X(x) > 0$. The *conditional density function* of $F_{Y|X}$, written $f_{Y|X}(\cdot|x)$, is given by

$$f_{Y|X}(y|x) = \frac{f_{X,Y}(x,y)}{f_X(x)} \tag{A.23}$$

for any x such that $f_X(x) > 0$.

Bayes's Theorem. The chain rule for density functions follows directly from the definition (A.23)

$$f_{X,Y}(x,y) = f_{Y|X}(y|x) f_X(x). \tag{A.24}$$

Quite analogously, we have

$$f_{X,Y}(x,y) = f_{X|Y}(x|y) f_Y(y). \tag{A.25}$$

Taking (A.24) and (A.25) together, we obtain *Bayes's theorem* for density functions

$$f_{X|Y}(x|y) = \frac{f_{Y|X}(y|x) f_X(x)}{\int_{-\infty}^{\infty} f_{Y|X}(y|x) f_X(x)\, dx}. \tag{A.26}$$

Independence. Continuous variables X and Y are called *independent*, denoted as $X \perp Y$, if $\{X \leq x\}$ and $\{Y \leq y\}$ are independent events for all $x, y \in \mathbb{R}$. It can be shown that X and Y are independent if and only if

$$F_{X,Y}(x,y) = F_X(x) F_Y(y) \quad \text{for all } x, y \in \mathbb{R},$$

or if and only if

$$f_{X,Y}(x,y) = f_X(x) f_Y(y) \tag{A.27}$$

whenever $F_{X,Y}$ is differentiable at (x,y) where $f_{X,Y}$, f_X, f_Y are taken to be the appropriate derivatives of $F_{X,Y}$, F_X, F_Y.

Clearly, X and Y are independent if and only if $f_{Y|X}(y|x) = f_Y(y)$ for any x such that $f_X(x) > 0$.

Continuous variables X and Y are *conditionally independent* given another continuous variable Z, denoted as $X \perp Y|Z$, if and only if

$$f_{X,Y|Z}(x,y|z) = f_{X|Z}(x|z) f_{Y|Z}(y|z) \quad \text{for all } x, y \in \mathbb{R} \tag{A.28}$$

for any Z such that $f_Z(z) > 0$.

Expectation. The *expectation* of a continuous variable X with density function f is

$$E(X) = \int_{-\infty}^{\infty} x f(x) \, dx$$

whenever this integral exists.

If X and $g(X)$ are continuous random variables, then

$$E(g(X)) = \int_{-\infty}^{\infty} g(x) f(x) dx. \tag{A.29}$$

The *conditional expectation* of Y given $X = x$ is

$$E(Y|X = x) = \int_{-\infty}^{\infty} y f_{Y|X}(y|x) \, dy. \tag{A.30}$$

The conditional expectation $E(Y|X)$ satisfies

$$E(E(Y|X)) = E(Y).$$

A.4 Normal Distribution

A notable example of a continuous distribution is the *normal* or *Gaussian* distribution.

Univariate Distribution. X has a (univariate) normal distribution, written $N(\mu, \sigma^2)$, if its density function is

$$f(x) = \frac{1}{\sqrt{2\pi\sigma^2}} \exp\left(-\frac{1}{2\sigma^2}(x-\mu)^2\right), \quad -\infty < x < \infty.$$

The mean and variance of the $N(\mu, \sigma^2)$ distribution are μ and σ^2, respectively.

Multivariate Distribution. $(X_1, \ldots, X_n)$ has a multivariate normal distribution, written $N(\mu, V)$, if its joint density function is

$$f(x) = (2\pi)^{-\frac{n}{2}} |V|^{-\frac{1}{2}} \exp\left(-\frac{1}{2}(x-\mu)^T V^{-1}(x-\mu)\right)$$

where μ is now a column vector of dimension n and V is a positive definite matrix of size (n, n).

The mean and covariance matrix of the $N(\mu, V)$ distribution are μ and V, respectively.

A.5 Hölder Inequality

Suppose that $p, q > 1$ and $p^{-1} + q^{-1} = 1$. Consider the function

$$f(t) = \frac{t^p}{p} + \frac{t^{-q}}{q}$$

for $t > 0$. It is easy to verify that the function $f(t)$ has a unique minimum at $t = 1$, therefore $f(t) \geq f(1) = 1$ for $t > 0$. Substituting

$$t = \frac{x^{1/q}}{y^{1/p}}$$

with

$$x = \frac{|X|}{\left(E|X^p|\right)^{1/p}}, \quad y = \frac{|Y|}{\left(E|Y^q|\right)^{1/q}},$$

we obtain after some manipulations

$$\frac{|X|^p}{pE|X^p|} + \frac{|Y|^q}{qE|Y^q|} \geq \frac{|XY|}{\left(E|X^p|\right)^{1/p}\left(E|Y^q|\right)^{1/q}}.$$

By taking expectations of both sides and using the assumption $p^{-1} + q^{-1} = 1$, we have

$$\boxed{E|XY| \leq \left(E|X^p|\right)^{1/p}\left(E|Y^q|\right)^{1/q}.} \tag{A.31}$$

Note that the equality in (A.31) holds if and only if $t = 1$, i.e.,

$$\frac{|X|^p}{E|X^p|} = \frac{|Y|^q}{E|Y^q|}. \tag{A.32}$$

B. Selected Topics from Convex Optimization

This section sums up only the very basic notions and results about optimization of convex functions. The propositions are stated without proofs, the reader interested in more background is referred to, e.g., Bazaraa and Shetty (1979), Avriel (1976) or Gill et al. (1981).

For Jensen's inequality see, e.g., Grimmett and Stirzaker (1992).

B.1 Convex Sets and Functions

Convex Set. A set $A \subset \mathbb{R}^n$ is called *convex* if for every $x, y \in A$ and $0 \leq \lambda \leq 1$,

$$\lambda x + (1 - \lambda) y \in A. \tag{B.1}$$

It follows by induction that A contains any convex combination of its points $x_i \in A$, $i = 1, \ldots, n$, i.e., for every $\lambda_i \geq 0$, $i = 1, \ldots, n$ such that $\sum_{i=1}^n \lambda_i = 1$, we have

$$\sum_{i=1}^n \lambda_i x_i \in A.$$

Convex Function. A function $f : \mathbb{R}^n \to \mathbb{R}$ is said to be *convex* on a convex set $A \subset \mathbb{R}^n$ if for every $x, y \in A$ and $0 \leq \lambda \leq 1$,

$$f(\lambda x + (1 - \lambda) y) \leq \lambda f(x) + (1 - \lambda) f(y). \tag{B.2}$$

A function f is said to be *strictly convex* if equality holds only for $\lambda = 0$ or $\lambda = 1$.

An alternative definition is as follows. A function $f : \mathbb{R}^n \to \mathbb{R}$ is called *convex* on a convex set $A \subset \mathbb{R}^n$ if for all $a \in \mathbb{R}^n$ there exists $\mu \in \mathbb{R}^n$, depending on a, such that

$$f(x) \geq f(a) + \sum_{i=1}^n \mu_i (x_i - a_i) \tag{B.3}$$

for all $x \in A$. A function f is said to be *strictly convex* if equality in (B.3) holds only for $x = a$.

Concave Function. A function f is concave if $-f$ is convex.

B.2 Minimization of Convex Functions

Necessary and Sufficient Condition for Minimum. Let $f: \mathbb{R}^n \to \mathbb{R}$ be a convex differentiable function. The point $x^* \in \mathbb{R}^n$ is a solution to the problem

$$\min_{x \in \mathbb{R}^n} f(x)$$

if and only if it satisfies the condition

$$\nabla_x f(x^*) = 0.$$

Existence of Minimum. Let $f(x)$ be convex on $\mathbb{R}^n$ and the set $\{x : f(x) \leq a\}$ be nonempty and bounded for some a. Then $f(x)$ has a minimum on $\mathbb{R}^n$.

Uniqueness of Minimum. Let $f(x)$ be strictly convex on $\mathbb{R}^n$. Suppose that x^* solves $\min_{x \in \mathbb{R}^n} f(x)$. Then x^* is the unique point of minimum.

B.3 Jensen's Inequality

Let $f : \mathbb{R}^n \to \mathbb{R}$ be convex and X be an n-dimensional random variable with finite mean. Setting $a = E(X)$ in (B.3), we have

$$f(X) \geq f(E(X)) + \sum_{i=1}^{n} \mu_i \left(x_i - E(X_i) \right)$$

for some fixed $\mu \in \mathbb{R}^n$. After taking expectation of both sides of the inequality, we obtain *Jensen's inequality*

$$E(f(X)) \geq f(E(X)). \tag{B.4}$$

If f is strictly convex, equality in (B.4) holds if and only if $X = E(X)$, i.e., X is a constant.

B.4 Legendre-Fenchel Transform

Let $f: \mathbb{R}^n \to \mathbb{R}$ be a convex function. The function

$$f^*(y) \overset{\Delta}{=} \sup_{x \in \mathbb{R}^n} \left(x^T y - f(x) \right) \tag{B.5}$$

for $y \in \mathbb{R}^n$ is called a *Legendre-Fenchel transform* of f or convex conjugate function to f.

If, in addition to being convex, the function f is also closed (continuous up to and including its boundary points), the Legendre-Fenchel transform of f has the following properties:

(a) f^* is a closed convex function on $\mathbb{R}^n$;

(b) $x^T y \leq f(x) + f^*(y)$ for all $x, y \in \mathbb{R}^n$;

(c) $f^{**} = f$, i.e., $f(x) = \sup_{y \in \mathbb{R}^n} \left(x^T y - f^*(y) \right)$.

See Bucklew (1990) or Ellis (1985) for more detail.

References

Akaike, H. (1973), Information theory and an extension of the maximum likelihood principle, *in* B. N. Petrov and F. Csáki (eds), *Proc. 2nd Internat. Symp. Information Theory*, Akadémiai Kiadó, Budapest, pp. 267–281.

Akaike, H. (1974), A new look at the statistical model identification, *IEEE Trans. Automat. Control* **19**, 716–723.

Ali, S. M. and Silvey, S. D. (1966), A general class of coefficients of divergence of one distribution from another, *J. Roy. Statist. Soc. Ser. B* **28**, 131–142.

Alspach, D. L. (1974), Gaussian sum approximations in nonlinear filtering and control, *Inform. Sci.* **7**, 271–290.

Amari, S. (1985), *Differential-Geometrical Methods in Statistics*, Vol. 28 of *Lecture Notes in Statistics*, Springer-Verlag, Berlin.

Anderson, B. D. O. and Moore, J. B. (1979), *Optimal Filtering*, Prentice-Hall, Englewood Cliffs, N.J.

Athans, M., Wishner, R. P. and Bertolini, A. (1968), Suboptimal state estimation for continuous-time nonlinear systems for discrete noisy measurements, *IEEE Trans. Automat. Control* **13**, 504–514.

Avriel, M. (1976), *Nonlinear Programming: Analysis and Methods*, Prentice-Hall, Englewood Cliffs, N.J.

Bahadur, R. R. (1954), Sufficiency and statistical decision functions, *Ann. Math. Statist.* **25**, 423–462.

Barndorff-Nielsen, O. E. (1978), *Information and Exponential Families in Statistical Theory*, Wiley, New York.

Barndorff-Nielsen, O. E., Cox, D. R. and Reid, N. (1986), The role of differential geometry in statistical theory, *Internat. Statist. Rev.* **54**, 83–96.

Barron, A. R. (1993), Universal approximation bounds for superposition of a sigmoidal function, *IEEE Trans. Inform. Theory* **39**, 930–945.

Barron, A. R. and Barron, R. L. (1988), Statistical learning networks: a unifying view, *Proceedings of the 20th Symposium on the Interface Between Computing Science and Statistics*, Alexandria, VA, pp. 192–203.

Barron, A. R. and Sheu, C. H. (1991), Approximation of density functions by sequences of exponential families, *Ann. Statist.* **19**, 1347–1369.

Bazaraa, M. S. and Shetty, C. M. (1979), *Nonlinear Programming: Theory and Algorithms*, Wiley, New York.

Beneš, V. (1981), Exact finite-dimensional filters for certain diffusions with nonlinear drift, *Stochastics* **5**, 65–92.

Beran, R. (1977), Minimum Hellinger distance estimates for parametric models, *Ann. Statist.* **5**, 445–463.

Bertsekas, D. and Shreve, S. E. (1978), *Stochastic Optimal Control: The Discrete Time Case*, Academic Press, New York.

Billings, S. A., Jamaluddin, H. B. and Chen, S. (1992), Properties of neural networks with applications to modelling non-linear dynamical systems, *Internat. J. Control* **55**, 193–224.

Billingsley, P. (1961), Statistical methods in Markov chains, *Ann. Math. Statist.* **32**, 12–40.

Blackwell, D. and Girshick, M. A. (1954), *Theory of Games and Statistical Decisions*, Wiley, New York.

Bohlin, T. (1970), Information pattern for linear discrete-time models with stochastic coefficients, *IEEE Trans. Automat. Control* **15**, 104–106.

Boltzmann, L. (1877), Beziehung zwischen dem zweiten Hauptsatze der mechanischen Wärmetheorie und der Wahrscheinlichkeitsrechnung respektive den Sätzen über das Wärmegleichgewicht, *Wien. Ber.* **76**, 373–435.

Boos, D. D. (1981), Minimum distance estimators for location and goodness of fit, *J. Amer. Statist. Assoc.* **76**, 663–670.

Boza, L. B. (1971), Asymptotically optimal tests for finite Markov chains, *Ann. Math. Statist.* **42**, 1992–2007.

Breiman, L., Friedman, J. H., Olshen, R. A. and Stone, C. J. (1984), *Classification and Regression Trees*, Wadsworth, Belmont, Calif.

Brigo, D. (1995), On the nice behaviour of the Gaussian Projection Filter with small observation noise, *Proceedings of the 3rd European Control Conference*, Rome, Italy, Vol. 3, pp. 1682–1687.

Brigo, D., Hanzon, B. and Gland, F. Le (1995), A differential-geometric approach to nonlinear filtering: the projection filter, *Proceedings of the 34th IEEE Conference on Decision and Control*, New Orleans, LA, Vol. 4, pp. 4006–4011.

Brockett, R. W. (1979), Classification and equivalence in estimation theory, *Proceedings of the 18th IEEE Conference on Decision and Control*, Ft. Lauderdale, FL, pp. 172–175.

Brockett, R. W. (1980), Remarks on finite-dimensional nonlinear estimation, *in* C. Lobry (ed.), *Analyse des systèmes*, Vol. 76 of *Astérisque*, Société mathématique de France, Paris.

Brockett, R. W. (1981), Nonlinear systems and nonlinear estimation theory, *in* Hazewinkel and Willems (1981), pp. 441–477.

Brockett, R. W. and Clark, J. M. C. (1980), The geometry of the conditional density equation, *in* O. L. R. Jacobs (ed.), *Analysis and Optimization of Stochastic Systems*, Academic Press, New York, pp. 299–309.

Brown, L. D. (1987), *Fundamentals of Statistical Exponential Families (with Applications in Statistical Decision Theory)*, Vol. 9 of *Lecture Notes — Monograph Series*, Inst. Math. Statist., Hayward, CA.

Bucklew, J. A. (1990), *Large Deviation Techniques in Decision, Simulation, and Estimation*, Wiley, New York.

Bucy, R. S. (1969), Bayes' theorem and digital realization for nonlinear filters, *J. Astronaut. Sci.* **17**, 80–94.

Bucy, R. S. and Senne, K. D. (1971), Digital synthesis of non-linear filters, *Automatica — J. IFAC* **7**, 287–298.

Burbea, J. and Rao, C. R. (1982), Entropy differential metric distance and divergence measures in probability spaces: a unified approach, *J. Multivariate Anal.* **12**, 575–596.

Burr, R. L. (1989), Iterative convex *I*-projection algorithms for maximum entropy and minimum cross-entropy computations, *IEEE Trans. Inform. Theory* **35**, 695–698.

Byrnes, C. I. and Lindquist, A. (eds) (1986), *Theory and Applications of Nonlinear Control Systems*, North Holland-Elsevier Science Publishers, Amsterdam.

Byrnes, C. I., Martin, C. F. and Saeks, R. E. (eds) (1988), *Analysis and Control of Nonlinear Systems*, North Holland-Elsevier Science Publishers, Amsterdam.

Campbell, L. L. (1985), The relation between information theory and the differential geometry approach to statistics, *Inform. Sci.* **35**, 199–210.

Carlin, B. P., Polson, N. G. and Stoffer, D. S. (1992), A Monte Carlo approach to nonnormal and nonlinear state-space modeling, *J. Amer. Statist. Assoc.* **87**, 493–500.

Čencov, N. N. (1964), The geometry of a "manifold" of probability distributions (in Russian), *Dokl. Akad. Nauk SSSR* **158**, 543–546. English translation in Soviet Math. Dokl. 5 (1964).

Čencov, N. N. (1966), Towards a systematic theory of exponential families of probability distributions (in Russian), *Teor. Verojatnost. i Primenen.* **11**, 483–494. English translation in Theor. Probability Appl. 11 (1966).

Čencov, N. N. (1972), *Statistical Decision Rules and Optimal Inference* (in Russian), Nauka, Moscow. English translation in *Translations of Mathematical Monographs* **53** (1982), Amer. Math. Soc., Providence, RI.

Center, J. L. (1971), Practical nonlinear filtering of discrete observations by generalized least-squares approximation of the conditional probability distribution, *Proceedings of the 2nd Symposium on Nonlinear Identification*, San Diego, CA, pp. 88–99.

Chen, S., Billings, S. A. and Grant, P. M. (1990), Non-linear systems identification using neural networks, *Internat. J. Control* **51**, 1191–1214.

Christensen, E. S. (1989), Statistical properties of *I*-projections within exponential families, *Scand. J. Statist.* **16**, 307–318.

Cover, T. M. and Thomas, J. A. (1991), *Elements of Information Theory*, Wiley, New York.

Cox, R. T. (1946), Probability, frequency and reasonable expectation, *Amer. J. Phys.* **14**, 1–13.

Csiszár, I. (1967), Information-type measures of difference of probability distributions and indirect observations, *Studia Sci. Math. Hungar.* **2**, 299–318.

Csiszár, I. (1975), *I*-divergence geometry of probability distributions and minimization problems, *Ann. Probab.* **3**, 146–158.

Csiszár, I. (1984), Sanov property, generalized *I*-projection and a conditional limit theorem, *Ann. Probab.* **12**, 768–793.

Csiszár, I. (1985), An extended maximum entropy principle and a Bayesian justification, *in* J. M. Bernardo, M. H. DeGroot, D. V. Lindley and A. F. M. Smith (eds), *Bayesian Statistics 2*, North Holland-Elsevier Science Publishers, Amsterdam, pp. 83–98.

Csiszár, I. and Körner, J. (1981), *Information Theory: Coding Theorems for Discrete Memoryless Systems*, Academic Press, New York.

Csiszár, I., Cover, T. M. and Choi, B.-S. (1987), Conditional limit theorem under Markov conditioning, *IEEE Trans. Inform. Theory* **33**, 788–801.

Darmois, G. (1935), Sur les lois de probabilité à estimation exhaustive, *C. R. Acad. Sci. Paris* **260**, 1265–1266.

Daum, F. E. (1988), New exact nonlinear filters, *in* J. C. Spall (ed.), *Bayesian Analysis of Time Series and Dynamic Models*, Marcel Dekker, New York, pp. 199–226.

Davies, L. and Gather, U. (1993), The identification of multiple outliers (with comments and rejoinder), *J. Amer. Statist. Assoc.*

Davis, M. H. A. and Marcus, S. I. (1981), An introduction to nonlinear filtering, *in* Hazewinkel and Willems (1981), pp. 53–75.

Davis, M. H. A. and Varaiya, P. (1972), Information states for linear stochastic systems, *J. Math. Anal. Appl.* **37**, 384–402.

Dawid, A. P. (1975), Discussion to the paper by B. Efron: Defining the curvature of a statistical problem (with application to second order efficiency), *Ann. Statist.* **3**, 1231–1234.

de Boor, C. and Rice, J. R. (1979), An adaptive algorithm for multivariate approximation giving optimal convergence rates, *J. Approx. Theory* **25**, 337–359.

de Figueiredo, R. J. P. and Jan, J. G. (1971), Spline filters, *Proceedings of the 2nd Symposium on Nonlinear Identification*, San Diego, CA, pp. 127–138.

de Finetti, B. (1937), Foresight: its logical laws, its subjective sources, *in* H. E. Kyburg and H. E. Smokler (eds), *Studies in Subjective Probability*, Wiley, New York, 1964, pp. 93–158.

de Finetti, B. (1974), *Theory of Probability; A Critical Introductory Treatment*, Vol. 1, Wiley, New York.

Donoho, D. L. and Johnstone, I. M. (1989), Projection-based approximation and a duality with kernel methods, *Ann. Statist.* **17**, 58–106.

Donsker, M. D. and Varadhan, S. R. S. (1975a), Asymptotic evaluation of certain Markov process expectations for large time, I., *Comm. Pure Appl. Math.* **28**, 1–47.

Donsker, M. D. and Varadhan, S. R. S. (1975b), Asymptotic evaluation of certain Markov process expectations for large time, II., *Comm. Pure Appl. Math.* **28**, 279–301.

Donsker, M. D. and Varadhan, S. R. S. (1976), Asymptotic evaluation of certain Markov process expectations for large time, III., *Comm. Pure Appl. Math.* **29**, 389–461.

Donsker, M. D. and Varadhan, S. R. S. (1983), Asymptotic evaluation of certain Markov process expectations for large time, IV., *Comm. Pure Appl. Math.* **36**, 183–212.

Dykstra, R. L. (1985), An iterative procedure for obtaining I-projections onto the intersection of convex sets, *Ann. Probab.*. **13**, 975–984.

Dynkin, E. B. (1951), Necessary and sufficient statistics for a family of probability distributions (in Russian), *Uspekhi matem. nauk* **VI**, 68–90.

Efron, B. (1975), Defining the curvature of a statistical problem (with applications to second order efficiency) (with discussion), *Ann. Statist.* **3**, 1189–1242.

Ellis, R. S. (1985), *Entropy, Large Deviations, and Statistical Mechanics*, Springer-Verlag, Berlin.

Fisher, R. A. (1922), On the mathematical foundations of theoretical statistics, *Roy. Soc. Phil. Trans. Ser. A* **222**, 309–368.

Fisher, R. A. (1925), Theory of statistical estimation, *Proc. Camb. Phil. Soc.* **22**, 700–725.

Fisher, R. A. (1934), Two new properties of mathematical likelihood, *Proc. R. Soc. London A* **144**, 285–307.

Friedman, J. H. (1991), Multivariate adaptive regression splines (with discussion), *Ann. Statist.* **19**, 1–141.

Friedman, J. H. and Stuetzle, W. (1981), Projection pursuit regression, *J. Amer. Statist. Assoc.* **76**, 817–823.

Gallant, A. R. (1987), *Nonlinear Statistical Models*, Wiley, New York.

Gelfand, A. E. and Smith, A. F. M. (1990), Sampling-based approaches to calculating marginal densities, *J. Amer. Statist. Assoc.* **85**, 398–409.

Gelfand, A. E. and Smith, A. F. M. (1991), Gibbs sampling for marginal posterior expectations, *Comm. Statist. A — Theory Methods* **20**, 1747–1766.

Geman, S. and Geman, D. (1984), Stochastic relaxation, Gibbs distributions, and the Bayesian restoration of images, *IEEE Trans. Pattern Anal. Machine Intell.* **PAMI-6**, 721–741.

Geman, S. and Hwang, C.-R. (1986), Diffusions for global optimization, *SIAM J. Control Optim.* **24**, 1031–1043.

Gill, P. E., Murray, W. and Wright, M. H. (1981), *Practical Optimization*, Academic Press, New York.

Gordon, N. J., Salmond, D. J. and Smith, A. F. M. (1993), A novel approach to nonlinear/non-Gaussian Bayesian state estimation, *Proc. IEE-F* **140**, 107–113.

Grimmett, G. R. and Stirzaker, D. R. (1992), *Probability and Random Processes*, second edn, Oxford University Press, Oxford.

Halmos, P. R. and Savage, L. J. (1949), Application of the Radon-Nikodym theorem to the theory of sufficient statistics, *Ann. Math. Statist.* **20**, 225–241.

Hampel, F. R., Ronchetti, E. M., Rousseeuw, P. J. and Stahel, W. A. (1986), *Robust Statistics: The Approach Based on Influence Functions*, Wiley, New York.

Hanzon, B. (1987), A differential-geometric approach to approximate nonlinear filtering, *in* C. T. J. Dodson (ed.), *Geometrization of Statistical Theory*, ULDM Publications, Lancaster, England, pp. 219–224.

Hanzon, B. and Hut, R. (1991), New results on the projection filter, *Proceedings of the 1st European Control Conference*, Grenoble, France, pp. 623–628.

Hartigan, J. A. (1967), The likelihood and invariance principles, *J. Roy. Statist. Soc. Ser. B* **29**, 533–539.

Hazewinkel, M. and Marcus, S. I. (1981), Some results and speculations on the role of Lie algebras in filtering, *in* Hazewinkel and Willems (1981), pp. 591–604.

Hazewinkel, M. and Marcus, S. I. (1982), On Lie algebras and finite-dimensional filtering, *Stochastics* **7**, 29–62.

Hazewinkel, M. and Willems, J. C. (eds) (1981), *Stochastic Systems: The Mathematics of Filtering and Identification and Applications*, D. Reidel, Dordrecht.

Heunis, A. J. (1990), On the stochastic differential equations of filtering theory, *Appl. Math. Comput.* **39**, 3s–36s.

Hoeffding, W. (1965), Asymptotically optimal tests for multinomial distributions, *Ann. Math. Statist.* **36**, 369–401.

Huber, P. J. (1981), *Robust Statistics*, Wiley, New York.

Ireland, C. T. and Kullback, S. (1968), Contingency tables with given marginals, *Biometrika* **55**, 179–188.

Isidori, A. (1989), *Nonlinear Control Systems: An Introduction*, Communications and Control Engineering Series, second edn, Springer-Verlag, Berlin.

Jacobs, O. L. R. (ed.) (1980), *Analysis and Optimization of Stochastic Systems*, Academic Press, New York.

Jaynes, E. T. (1957), Information theory and statistical mechanics, *Phys. Rev. A (3)* **106**, 620–630.

Jaynes, E. T. (1979), Where do we stand on maximum entropy?, *in* R. O. Levine and M. Tribus (eds), *The Maximum Entropy Formalism*, MIT Press, Cambridge, MA, pp. 15–118.

Jazwinski, A. H. (1966), Filtering of nonlinear systems, *IEEE Trans. Automat. Control* **11**, 765–766.

Jazwinski, A. H. (1970), *Stochastic Processes and Filtering Theory*, Academic Press, New York.

Jeffreys, H. (1939), *Theory of Probability*, Oxford University Press, Oxford.

Jones, L. K. (1989), Approximation-theoretic derivation of logarithmic entropy principles for inverse problems and unique extension of the maximum entropy method to incorporate prior knowledge, *SIAM J. Appl. Math.* **49**, 650–661.

Jones, L. K. and Byrne, C. L. (1990), General entropy criteria for inverse problems, with applications to data compression, pattern classification, and cluster analysis, *IEEE Trans. Inform. Theory* **36**, 23–30.

Juditsky, A., Hjalmarsson, H., Benveniste, A., Delyon, B., Ljung, L., Sjöberg, J. and Zhang, Qinghua (1985), Nonlinear black-box modeling in system identification: mathematical foundations, *Automatica — J. IFAC* **31**, 1725–1750.

Justesen, J. and Høholdt, T. (1984), Maxentropic Markov chains, *IEEE Trans. Inform. Theory* **30**, 665–667.

Kalman, R. E. (1960), A new approach to linear filtering and prediction problem, *Trans. ASME, Ser. D, J. Basic Engin.* **82**, 35–45.

Kannappan, Pl. (1972a), On directed divergence and inaccuracy, *Z. Wahrsch. Verw. Gebiete* **25**, 49–55.

Kannappan, Pl. (1972b), On Shannon's entropy, directed divergence and inaccuracy, *Z. Wahrsch. Verw. Gebiete* **22**, 95–100.

Kárný, M., Halousková, A., Böhm, J., Kulhavý, R. and Nedoma, P. (1985), Design of linear quadratic adaptive control: Theory and algorithms for practice, *Kybernetika*. Supplement to No. 3, 4 ,5, 6.

Kass, R. E., Tierney, L. and Kadane, J. B. (1988), Asymptotics in Bayesian computation, *in* J. M. Bernardo, M. H. DeGroot, D. V. Lindley and A. F. M. Smith (eds), *Bayesian Statistics 3*, Clarendon Press, Oxford, pp. 261–278.

Kerridge, D. F. (1961), Inaccuracy and inference, *J. Roy. Statist. Soc. Ser. B* **23**, 284–294.

Keynes, J. M. (1921), *A Treatise on Probability*, MacMillan, London.

Khintchine, A. Ja. (1943), *Mathematical Foundations of Statistical Mechanics*, OGIZ, Moscow. English translation in Dover, New York, 1949.

Kolmogorov, A. N. (1942), Definitions of center of dispersion and measure of accuracy from a finite number of observations, *Izv. Akad. Nauk SSSR Ser. Mat.* **6**, 3–32.

Koopman, B. O. (1936), On distributions admitting a sufficient statistic, *Trans. Amer. Math. Soc.* **39**, 399–409.

Kramer, S. C. and Sorenson, H. W. (1988), Recursive Bayesian estimation using piece-wise constant approximations, *Automatica — J. IFAC* **24**, 789–801.

Kriz, T. A. and Talacko, J. V. (1968), Equivalence of the maximum likelihood estimator to a minimum entropy estimator, *Trabajos Estadíst.* **19**, 55–65.

Kubo, R. (1965), *Statistical Mechanics*, North-Holland, Amsterdam.

Kulhavý, R. (1990), Recursive nonlinear estimation: a geometric approach, *Automatica — J. IFAC* **26**, 545–555.

Kulhavý, R. (1992), Recursive nonlinear estimation: geometry of a space of posterior densities, *Automatica — J. IFAC* **28**, 313–323.

Kulhavý, R. (1993), Can approximate Bayesian estimation be consistent with the ideal solution?, *Proceedings of the 12th IFAC World Congress*, Sydney, Australia, Vol. 4, pp. 225–228.

Kulhavý, R. (1994a), Bayesian estimation, large deviations, and incomplete data, *IEEE Conference on Decision and Control*, Orlando, FL, Vol. 1, pp. 755–756.

Kulhavý, R. (1994b), Can we preserve the structure of recursive Bayesian estimation in a limited-dimensional implementation?, *in* U. Helmke, R. Mennicken and J. Saurer (eds), *Systems and Networks: Mathematical Theory and Applications*, Vol. I, Akademie Verlag, Berlin, pp. 251–272.

Kulhavý, R. (1995a), A geometric approach to statistical estimation, *Proceedings of the 34th IEEE Conference on Decision and Control*, New Orleans, LA, Vol. 2, pp. 1097–1102.

Kulhavý, R. (1995b), A Kullback-Leibler distance approach to system identification, *Preprints of the IFAC Symposium on Adaptive Systems in Control and Signal Processing*, Budapest, Hungary, pp. 55–66.

Kulhavý, R. and Hrnčíř, F. (1994), Approximation and uncertainty in parameter estimation, *Preprints of the European IEEE Workshop on Computer-intensive Methods in Control and Signal Processing*, Prague, Czech Republic, pp. 61–70.

Kulhavý, R. and Kraus, F. J. (1996), On duality of regularized exponential and linear forgetting, *Proceedings of the 13th IFAC World Congress*, San Francisco, CA.

Kulhavý, R. and Zarrop, M. B. (1993), On a general concept of forgetting, *Internat. J. Control* **58**, 905–924.

Kullback, S. (1959), *Information Theory and Statistics*, Wiley, New York.

Kullback, S. (1968), Probability densities with given marginals, *Ann. Math. Statist.* **39**, 1236–1243.

Kullback, S. and Leibler, R. A. (1951), On information and sufficiency, *Ann. Math. Statist.* **22**, 79–86.

Kumar, P. R. (1985), A survey of some results in stochastic adaptive control, *SIAM J. Control Optim.* **23**, 329–380.

Kumar, P. R. and Varaiya, P. (1986), *Stochastic Systems: Estimation, Identification, and Adaptive Control*, Prentice-Hall, Englewood Cliffs, N.J.

Lauritzen, S. L. (1988), *Extremal Families and Systems of Sufficient Statistics*, Vol. 49 of *Lecture Notes in Statistics*, Springer-Verlag, Berlin.

Lehmann, E. L. (1959), *Testing Statistical Hypotheses*, Wiley, New York.

Lehmann, E. L. (1983), *Theory of Point Estimation*, Wiley, New York.

Lehmann, E. L. and Scheffé, H. (1950), Completeness, similar regions and unbiased estimation, *Sankhyā* **10**, 305–340.

Lehmann, E. L. and Scheffé, H. (1955), Completeness, similar regions and unbiased estimation, *Sankhyā* **15**, 219–236.

Li, M. and Vitanyi, P. M. B. (1993), *An Introduction to Kolmogorov Complexity and Its Applications*, Springer-Verlag, Berlin.

Light, W. A. and Cheney, E. W. (1985), *Approximation Theory in Tensor Product Spaces*, Springer-Verlag, Berlin.

Lindgren, B. W. (1976), *Statistical Theory*, third edn, MacMillan, New York.

Lindley, D. V. (1980), Approximate Bayesian methods, *in* J. M. Bernardo, M. H. DeGroot, D. V. Lindley and A. F. M. Smith (eds), *Bayesian Statistics 1*, University Press, Valencia, pp. 221–237.

Linnik, Ju. V. (1966), *Statistical Problems with Nuisance Parameters*, Nauka, Moscow. English translation in Amer. Math. Soc., Providence, RI, 1968.

Linnik, Ju. V. (1967), *Leçons sur les problèmes de statistique analytique*, Gauthier-Villars, Paris.

Ljung, L. (1987), *System Identification: Theory for the User*, Prentice-Hall, Englewood Cliffs, N.J.

Masi, G. B. Di and Taylor, T. J. (1991), A new approximation method for nonlinear filtering using nilpotent harmonic analysis, *Proceedings of the 30th IEEE Conference on Decision and Control*, Brighton, England, pp. 2750–2751.

Matsuoka, T. and Ulrych, T. J. (1986), Information theory measures with application to model identification, *IEEE Trans. Acoust., Speech, Signal Processing* **34**, 511–517.

Metropolis, N., Rosenbluth, A. W., Rosenbluth, M. N., Teller, A. H. and Teller, E. (1953), Equations of state calculations by fast computing machines, *J. Chem. Phys.* **21**, 1087–1092.

Narendra, K. S. and Parthasarathy, K. (1990), Identification and control of dynamical systems using neural networks, *IEEE Trans. Neural Networks* **1**, 4–27.

Natarajan, S. (1985), Large deviations, hypotheses testing, and source coding for finite Markov chains, *IEEE Trans. Inform. Theory* **31**, 360–365.

Naylor, J. C. and Smith, A. F. M. (1988), Econometric illustrations of novel numerical integration strategies for Bayesian inference, *J. Econometrics* **38**, 103–125.

Nevel'son, M. B. and Has'minskiĭ, R. Z. (1973), *Stochastic Approximation and Recursive Estimation*, American Mathematical Society, Providence, RI.

Neyman, J. (1935), Su un teorema concernente le cosiddette statistiche sufficienti, *Inst. Ital. Atti. Giorn.* **6**, 320–334.

Nijmeijer, H. and van der Schaft, A. J. (1990), *Nonlinear Dynamical Control Systems*, Springer-Verlag, New York.

Ocone, D. L. (1981), Finite dimensional estimation algebras in nonlinear filtering, *in* Hazewinkel and Willems (1981), pp. 629–636.

O'Sullivan, J. A., Miller, M. I., Srivastava, A. and Snyder, D. L. (1993), Tracking using a random sampling algorithm, *Proceedings of the 12th IFAC World Congress*, Sydney, Australia, Vol. 5, pp. 435–438.

Papoulis, A. (1991), *Probability, Random Variables, and Stochastic Processes*, third edn, McGraw-Hill, New York.

Parr, W. C. and Schucany, W. R. (1980), Minimum distance and robust estimation, *J. Amer. Statist. Assoc.* **75**, 616–624.

Pázman, A. (1993), *Nonlinear Statistical Models*, Kluwer, Dordrecht.

Peeters, R. (1994), *System Identification Based on Riemannian Geometry: Theory and Algorithms*, Thesis Publishers, Amsterdam.

Peterka, V. (1981), Bayesian approach to system identification, *in* P. Eykhoff (ed.), *Trends and Progress in System Identification*, Pergamon, Elmsford, N.Y., chapter 8, pp. 239–304.

Pitman, E. J. G. (1936), Sufficient statistics and intrinsic accuracy, *Proc. Camb. Phil. Soc.* **32**, 567–579.

Poggio, T. and Girosi, F. (1990), Networks for approximation and learning, *Proc. IEEE* **78**, 1481–1497.

Rao, C. R. (1945), Information and accuracy attainable in the estimation of statistical parameters, *Bull. Calcutta Math. Soc.* **37**, 81–91.

Ripley, B. (1987), *Stochastic Simulation*, Wiley, New York.

Rissannen, J. (1989), *Stochastic Complexity in Statistical Inquiry*, World Scientific, Singapore.

Robbins, H. (1956), An empirical Bayes approach to statistics, *Proceedings of the Third Berkeley Symposium on Mathematical Statistics and Probability*, University of California Press, Berkeley, CA, Vol. I, pp. 157–163.

Robbins, H. (1964), The empirical Bayes approach to statistical decision problems, *Ann. Math. Statist.* **35**, 1–20.

Robert, C. P. (1989), *The Bayesian Choice: A Decision-Theoretic Motivation*, Springer-Verlag, Berlin.

Runggaldier, W. J. and Visentin, C. (1990), Combined filtering and parameter estimation: approximations and robustness, *Automatica — J. IFAC* **26**, 401–404.

Sanner, R. M. and Slotine, J.-J. E. (1991), Stable adaptive control and recursive identification using radial Gaussian networks, *Proceedings of the 30th IEEE Conference on Decision and Control*, Brighton, England.

Sanov, I. N. (1957), On the probability of large deviations of random variables (in Russian), *Mat. Sb. (N.S.)* **42**, 11–44. English translation in *Sel. Transl. Math. Statist. Probab.* **I** (1961), 213–244.

Savage, L. J. (1954), *The Foundations of Statistics*, Wiley, New York.

Shannon, C. E. (1948), A mathematical theory of communication, *Bell System Tech. J.* **26**, 379–423, 623–656.

Shiryaev, A. N. (1984), *Probability*, Springer-Verlag, New York.

Shore, J. E. (1981), Minimum cross-entropy spectral analysis, *IEEE Trans. Acoust., Speech, Signal Processing* **29**, 230–237.

Shore, J. E. and Johnson, R. W. (1980), Axiomatic derivation of the principle of maximum entropy and the principle of minimum cross entropy, *IEEE Trans. Inform. Theory* **26**, 26–37.

Silverman, B. W. (1986), *Density Estimation for Statistics and Data Analysis*, Chapman & Hall, London.

Sjöberg, J. (1995), *Non-linear system identification with neural networks*, Linköping Studies in Science and Technology, Dissertations, no. 381, Linköping University.

Sjöberg, J., Zhang, Qinghua, Ljung, L., Benveniste, A., Delyon, B., Glorennec, P.-Y., Hjalmarsson, H. and Juditsky, A. (1985), Nonlinear black-box modeling in system identification: a unified overview, *Automatica — J. IFAC* **31**, 1691–1724.

Smith, A. F. M. and Gelfand, A. E. (1992), Bayesian statistics without tears: a sampling-resampling perspective, *Amer. Statist.* **46**, 84–88.

Smith, A. F. M., Skene, A. M., Shaw, J. E. H., Naylor, J. C. and Dransfield, M. (1987), Progress with numerical and graphical methods for practical Bayesian statistics, *The Statistician* **36**, 75–82.

Sorenson, H. W. (1966), Kalman filtering techniques, *in* C. T. Leondes (ed.), *Advances in Control Systems*, Vol. 3, Academic Press, New York, pp. 63–80.

Sorenson, H. W. (1974), On the development of practical nonlinear filters, *Inform. Sci.* **7**, 253–270.

Sorenson, H. W. (1988), Recursive estimation for nonlinear dynamic systems, *in* J. C. Spall (ed.), *Bayesian Analysis of Time Series and Dynamic Models*, Marcel Dekker, New York, pp. 127–165.

Sorenson, H. W. and Alspach, D. L. (1971), Recursive Bayesian estimation using Gaussian sums, *Automatica — J. IFAC* **7**, 465–479.

Sorenson, H. W. and Stubberud, A. R. (1968), Recursive filtering for systems with small but nonnegligible nonlinearities, *Internat. J. Control* **7**, 271–280.

Spitzer, F. (1972), A variational characterization of finite Markov chains, *Ann. Math. Statist.* **43**, 580–583.

Srinivasan, K. (1970), State estimation by orthogonal expansion of probability distributions, *IEEE Trans. Automat. Control* **15**, 3–10.

Stoorvogel, A. A. and van Schuppen, J. H. (1995), System identification with information theoretic criteria, *Report BS-R9513*, CWI, Amsterdam.

Striebel, C. (1982), Approximate finite dimensional filters for some nonlinear problems, *Stochastics* **7**, 183–203.

Tanner, M. A. and Wong, W. H. (1987), The calculation of posterior distributions by data augmentation (with discussion), *J. Amer. Statist. Assoc.* **82**, 528–547.

Tierney, L. and Kadane, J. B. (1986), Accurate approximations for posterior moments and marginal densities, *J. Amer. Statist. Assoc.* **81**, 82–86.

Tierney, L., Kass, R. E. and Kadane, J. B. (1989a), Approximate marginal densities of nonlinear functions, *Biometrika* **76**, 425–433.

Tierney, L., Kass, R. E. and Kadane, J. B. (1989b), Fully exponential Laplace approximations to expectations and variances of nonpositive functions, *J. Amer. Statist. Assoc.* **84**, 710–716.

Vajda, I. (1982), A new general approach to minimum distance estimation, *Transactions of the Ninth Prague Conference on Information Theory, Statistical Decision Functions, Random Processes*, Academia, Prague.

Vajda, I. (1984a), Asymptotic efficiency and robustness of *D*-estimators, *Kybernetika (Prague)* **20**, 358–375.

Vajda, I. (1984b), Consistency of *D*-estimators, *Kybernetika (Prague)* **20**, 283–303.

Vajda, I. (1984c), Motivation, existence and equivariance of *D*-estimators, *Kybernetika (Prague)* **20**, 189–208.

Vajda, I. (1989), *Theory of Statistical Inference and Information*, Kluwer, Dordrecht.

van Campenhout, J. M. and Cover, T. M. (1981), Maximum entropy and conditional probability, *IEEE Trans. Inform. Theory* **27**, 483–489.

Wang, A. H. and Klein, R. L. (1976), Implementation of nonlinear estimators using monospline, *Proceedings of the 13th IEEE Conference on Decision and Control*, pp. 1305–1307.

Weisberg, S. (1980), *Applied Linear Regression*, Wiley, New York.

Whittle, P. (1955), Some distributions and moment formulae for the Markov chain, *J. Roy. Statist. Soc. Ser. B* **17**, 235–242.

Wiberg, D. M. (1987), Another approach to on-line parameter estimation, *Proceedings of the 1987 American Control Conference*, Minneapolis, MN, Vol. 1, pp. 418–423.

Wiberg, D. M. and DeWolf, D. G. (1991), A convergent approximation of the optimal parameter estimator, *Proceedings of the 30th IEEE Conference on Decision and Control*, Brighton, England, pp. 2017–2023.

Willems, J. C. (1980), Some remarks on the concept of information state, *in* O. L. R. Jacobs (ed.), *Analysis and Optimization of Stochastic Systems*, Academic Press, New York, pp. 285–295.

Wolfowitz, J. (1957), The minimum distance method, *Ann. Math. Statist.* **28**, 75–88.

Zacks, S. (1971), *The Theory of Statistical Inference*, Wiley, New York.

Index

Lecture Notes in Control and Information Sciences

Edited by M. Thoma

Vol. 180: Kall, P. (Ed.)
System Modelling and Optimization.
Proceedings of the 15th IFIP Conference,
Zurich, Switzerland, September 2-6, 1991
969 pp. 1992 [3-540-55577-3]

Vol. 181: Drane, C.R.
Positioning Systems - A Unified Approach
168 pp. 1992 [3-540-55850-0]

Vol. 182: Hagenauer, J. (Ed.)
Advanced Methods for Satellite and Deep
Space Communications. Proceedings of
an International Seminar Organized by
Deutsche Forschungsanstalt für Luft-und
Raumfahrt (DLR), Bonn, Germany,
September 1992
196 pp. 1992 [3-540-55851-9]

Vol. 183: Hosoe, S. (Ed.)
Robust Control. Proceesings of a Workshop
held in Tokyo, Japan, June 23-24, 1991
225 pp. 1992 [3-540-55961-2]

Vol. 184: Duncan, T.E.; Pasik-Duncan, B.
(Eds)
Stochastic Theory and Adaptive Control.
Proceedings of a Workshop held in
Lawrence, Kansas, September 26-28,
1991
500 pp. 1992 [3-540-55962-0]

Vol. 185: Curtain, R.F. (Ed.); Bensoussan,
A.; Lions, J.L.(Honorary Eds)
Analysis and Optimization of Systems:
State and Frequency Domain Approaches
for Infinite-Dimensional Systems.
Proceedings of the 10th International
Conference, Sophia-Antipolis, France, June
9-12, 1992.
648 pp. 1993 [3-540-56155-2]

Vol. 186: Sreenath, N.
Systems Representation of Global Climate
Change Models. Foundation for a Systems
Science Approach.
288 pp. 1993 [3-540-19824-5]

Vol. 187: Morecki, A.; Bianchi, G.;
Jaworeck, K. (Eds)
RoManSy 9: Proceedings of the Ninth
CISM-IFToMM Symposium on Theory and
Practice of Robots and Manipulators.
476 pp. 1993 [3-540-19834-2]

Vol. 188: Naidu, D. Subbaram
Aeroassisted Orbital Transfer: Guidance
and Control Strategies
192 pp. 1993 [3-540-19819-9]

Vol. 189: Ilchmann, A.
Non-Identifier-Based High-Gain Adaptive
Control
220 pp. 1993 [3-540-19845-8]

Vol. 190: Chatila, R.; Hirzinger, G. (Eds)
Experimental Robotics II: The 2nd
International Symposium, Toulouse,
France, June 25-27 1991
580 pp. 1993 [3-540-19851-2]

Vol. 191: Blondel, V.
Simultaneous Stabilization of Linear
Systems
212 pp. 1993 [3-540-19862-8]

Vol. 192: Smith, R.S.; Dahleh, M. (Eds)
The Modeling of Uncertainty in Control
Systems
412 pp. 1993 [3-540-19870-9]

Vol. 193: Zinober, A.S.I. (Ed.)
Variable Structure and Lyapunov Control
428 pp. 1993 [3-540-19869-5]

Vol. 194: Cao, Xi-Ren
Realization Probabilities: The Dynamics of
Queuing Systems
336 pp. 1993 [3-540-19872-5]

Vol. 195: Liu, D.; Michel, A.N.
Dynamical Systems with Saturation
Nonlinearities: Analysis and Design
212 pp. 1994 [3-540-19888-1]

Vol. 196: Battilotti, S.
Noninteracting Control with Stability for
Nonlinear Systems
196 pp. 1994 [3-540-19891-1]

Vol. 197: Henry, J.; Yvon, J.P. (Eds)
System Modelling and Optimization
975 pp approx. 1994 [3-540-19893-8]

Vol. 198: Winter, H.; Nüßer, H.-G. (Eds)
Advanced Technologies for Air Traffic Flow
Management
225 pp approx. 1994 [3-540-19895-4]

Vol. 199: Cohen, G.; Quadrat, J.-P. (Eds)
11th International Conference on
Analysis and Optimization of Systems –
Discrete Event Systems: Sophia-Antipolis,
June 15–16–17, 1994
648 pp. 1994 [3-540-19896-2]

Vol. 200: Yoshikawa, T.; Miyazaki, F. (Eds)
Experimental Robotics III: The 3rd
International Symposium, Kyoto, Japan,
October 28-30, 1993
624 pp. 1994 [3-540-19905-5]

Vol. 201: Kogan, J.
Robust Stability and Convexity
192 pp. 1994 [3-540-19919-5]

Vol. 202: Francis, B.A.; Tannenbaum, A.R.
(Eds)
Feedback Control, Nonlinear Systems,
and Complexity
288 pp. 1995 [3-540-19943-8]

Vol. 203: Popkov, Y.S.
Macrosystems Theory and its Applications:
Equilibrium Models
344 pp. 1995 [3-540-19955-1]

Vol. 204: Takahashi, S.; Takahara, Y.
Logical Approach to Systems Theory
192 pp. 1995 [3-540-19956-X]

Vol. 205: Kotta, U.
Inversion Method in the Discrete-time
Nonlinear Control Systems Synthesis
Problems
168 pp. 1995 [3-540-19966-7]

Vol. 206: Aganovic, Z.;.Gajic, Z.
Linear Optimal Control of Bilinear Systems
with Applications to Singular Perturbations
and Weak Coupling
133 pp. 1995 [3-540-19976-4]

Vol. 207: Gabasov, R.; Kirillova, F.M.;
Prischepova, S.V.
Optimal Feedback Control
224 pp. 1995 [3-540-19991-8]

Vol. 208: Khalil, H.K.; Chow, J.H.;
Ioannou, P.A. (Eds)
Proceedings of Workshop on Advances in
Control and its Applications
300 pp. 1995 [3-540-19993-4]

Vol. 209: Foias, C.; Özbay, H.;
Tannenbaum, A.
Robust Control of Infinite Dimensional
Systems: Frequency Domain Methods
230 pp. 1995 [3-540-19994-2]

Vol. 210: De Wilde, P.
Neural Network Models: An Analysis
164 pp. 1996 [3-540-19995-0]

Vol. 211: Gawronski, W.
Balanced Control of Flexible Structures
280 pp. 1996 [3-540-76017-2]

Vol. 212: Sanchez, A.
Formal Specification and Synthesis of
Procedural Controllers for Process Systems
248 pp. 1996 [3-540-76021-0]

Vol. 213: Patra, A.; Rao, G.P.
General Hybrid Orthogonal Functions and
their Applications in Systems and Control
144 pp. 1996 [3-540-76039-3]

Vol. 214: Yin, G.; Zhang, Q. (Eds)
Recent Advances in Control and
Optimization of Manufacturing Systems
240 pp. 1996 [3-540-76055-5]

Vol. 215: Bonivento, C.; Marro, G.;
Zanasi, R. (Eds)
Colloquium on Automatic Control
240 pp. 1996 [3-540-76060-1]